Computational Molecular Science and Engineering Forum

Presentations at the 2008 AIChE Annual Meeting

Philadelphia, Pennsylvania
16 - 21 November 2008

ISBN: 978-1-61567-233-2

Printed from e-media with permission by:

Curran Associates, Inc.
57 Morehouse Lane
Red Hook, NY 12571

CURRAN ASSOCIATES INC.
proceedings.com

Some format issues inherent in the e-media version may also appear in this print version.

Printed by Curran Associates, Inc. (2009)

AIChE
3 Park Avenue
New York, NY 10016-5991

Phone: (203) 702-7660
Fax: (203) 775-5177

www.aiche.org

Additional copies of this publication are available from:

Curran Associates, Inc.
57 Morehouse Lane
Red Hook, NY 12571 USA
Phone: 845-758-0400
Fax: 845-758-2634
Email: curran@proceedings.com
Web: www.proceedings.com

TABLE OF CONTENTS

185a Density Functional Study of Hydrogen Spillover on Pt and Pd Decorated MoO3

Liang Chen, Institute of Materials Technology and Engineering, Chinese Academy of Sciences, 519 Zhuangshi Blvd. C210, Ningbo, China

Hydrogen spillover on MoO_3 (010) surface in the presence of a platinum and palladium catalyst was modeled using periodic density functional theory (DFT). The migration of H from a saturated Pt_6 cluster to MoO_3 (010) surface was found to undergo a transition from repulsive electrostatic to attractive proton-oxygen interactions. The hydrogen is able to move nearly freely in the interface and diffuse into the bulk lattice at ambient temperatures. We show that the high mobility is largely attributed to the massive H-bonding network in the MoO_3 lattice.

185b Density Functional Theory Methods for Modeling Electrocatalysis: Application to Borohydride Oxidation Over Au and Pt(111)

Gholamreza Rostamikia and **Michael J. Janik**. Chemical Engineering, Pennsylvania State University, 6E Fenske Laboratory, University Park, PA 16802

Difficulties associated with experimental characterization of elementary kinetics for complex electrocatalytic reactions, such as the eight electron oxidation of borohydride species, motivates the application of density functional theory (DFT) methods to mechanism determination. A series of quantum-mechanical model system configurations will be presented, involving sequential addition of the complexities of the electrode-electrolyte interface, and the utility of each in evaluating elementary reaction energetics of both redox and chemical reaction steps will be discussed.

In this study, multiple approaches are discussed to approximate the electrochemical nature of the catalytic surface with the DFT model. The first method uses a metal-vacuum slab interface to determine the interaction of individual species with the electrode surface. The energy of ion or electron species produced or consumed in individual reaction steps is dictated by their energy in the bulk electrolyte or by the electrode potential. This method approximates that ion adsorption occurs with integral electron transfer and neglects solvation of surface bound species and the potential dependent interaction of surface species with an interfacial electric field. Corrections for dipole-field and solvation interactions may be added to energetics determined with the vacuum-slab model. The second method includes solvation and electric field variation within the model system in determining the energy of adsorbed species. This is done at increased cost in computational resources, and is used for a subset of calculations performed.

The ion adsorption energies as well as energetics and activation barriers of initial steps of borohydride oxidation over Au(111) and Pt(111) surfaces was calculated and will be discussed. Specifically, comparison between "vacuum slab" and solvated, variable potential methods in calculating ion adsorption equilibrium constants and surface reaction activation barriers will be presented.

185c Surface Poisoning of $RuO_2(110)$ by Formation of Carbonate

Hangyao Wang, Chemical and Biomolecular Engineering, University of Notre Dame, 182 Fitzpatrick Hall, Notre Dame, IN 46556 and William F. Schneider, Department of Chemical and Biomolecular Engineering, University of Notre Dame, 182 Fitzpatrick Hall, Notre Dame, IN 46556.

Base metal oxides have long been of interest as catalysts for oxidation of small molecules such as CO, NO, or SO_2, but practical applications are limited by surface poisoning.[1] With growing interest in the role of metal oxide films in the oxidation activity of even the more noble metals, it is important to understand and ultimately to learn to bypass surface poisoning. RuO_2 has been widely studied and characterized as a CO oxidation catalyst, and experimental observations suggest that this activity can be poisoned under some conditions by the formation of a surface carbonate.[2] In this work, we use plane-wave, supercell DFT calculations to explore the formation of carbonate on an $RuO_2(110)$ surface and determine its implications for surface poisoning. We show that several carbonate forms are possible. In general, carbonate adsorbs weakly and, based on equilibrium calculations, competes ineffectively with oxygen adsorption. The more favorable kinetics of carbonate formation are likely important to the observed high carbonate coverages [3] at high O_2 pressures.

[1]: K. Kolasinski, Surface Science: Foundations of Catalysis and Nanoscience, Wiley, (2002).

[2]: A. Lafosse, Y. Wang, and K. Jacobi, J. Chem. Phys., 117, 2823 (2002).

[3]: M. Rossler, S. Giinther, and J. Wintterlin, J. Phys. Chem. C, 111, 2242 (2007).

185d Ceria-Based Oxides for Catalysis Applications Investigated by the DFT +U Approach

Adam D. Mayernick, Department of Chemical Engineering, Pennsylvania State University, 133 Fenske Lab, University Park, PA 16802 and Michael Janik, Chemical Engineering, Pennsylvania State University, 158 Fenske Laboratory, University Park, PA 16802.

Ceria (CeO_2) offers unique properties as a heterogeneous catalyst or catalyst support for a number of applications, such as three-way automotive catalysts, preferential CO oxidation, and catalytic oxidation of hydrocarbons. For each of these applications, the use of ceria is motivated by its ability to store and release oxygen, or more generally to readily transition between oxidation states. The addition of low levels of rare-earth or noble metals alters the redox properties and hydrocarbon oxidation activity of ceria-based materials. We use density functional theory, with the inclusion of an on-site Coulombic interaction (DFT+U), to examine the energetics of oxygen vacancy formation and methane oxidation over ceria surfaces with the addition of Zr or Pd. We address the thermodynamic preference for Zr or Pd to substitute into the CeO_2 lattice, with specific emphasis on the relative stability of different phases as a function of temperature and oxidizing conditions. The thermodynamic preference for Zr or Pd to incorporate into the ceria lattice is evaluated by considering the relative energy of different phase combinations including segregated CeO_2 and bulk Zr or Pd phases, segregated CeO_2 and ZrO_2 or PdO phases, Zr or Pd incorporated into a $M_xCe_{1-x}O_2$ phase. The relative energy of each configuration is calculated as a function of partial pressure of oxygen (P_{O2}) and temperature. The incorporation of Zr or Pd into the CeO_2 lattice becomes thermodynamically favorable over segregated configurations at certain temperatures and oxygen partial pressures of relevance for hydrocarbon oxidation. The sensitivity of results to the value of the Hubbard U-term is discussed to rationalize the U value used herein, and to explore the use of DFT+U to accurately represent the electronic structure of ceria-based oxides.

The partial reduction of ceria may occur through the formation of oxygen vacancies, during which cerium atoms are formally reduced from Ce^{4+} to Ce^{3+}. The inclusion of the Hubbard U-term within the DFT approach corrects for the fact that DFT representations overestimate the preference for Ce 4f electrons to delocalize upon Ce reduction. This reduction results in partial occupation of 4f states, which is realized through the appearance of a "gap state" between the valence band and Ce 4f band in CeO_2. The value of U dictates the position of the gap state, and we calculate the energy gap between this state and the valence band for a series of U values. Oxygen vacancy formation in pure ceria and Zr-substituted ceria results in the reduction of cerium atoms, and we find a linear correlation between vacancy formation energy and U value. However, oxygen vacancy formation in Pd-substituted ceria reduces Pd rather than Ce, and thus the vacancy formation energy is weakly dependent on U value. The adsorption energy of methane, which also leads to reduction of the oxide surface, is more exothermic over Zr-substituted ceria surfaces than over pure ceria, and even more exothermic over Pd-substituted surfaces. These results provide insight into the mechanism by which metal addition alters the redox properties and catalytic activity of ceria. The energetics of reductive process on ceria can be influenced by the value of the Hubbard U-term, which allows for reasonable representations of the electronic structure. These results aid in both interpreting experimental behavior and guiding design of improved ceria-based catalysts.

185e Adsorption, Absorption, and Diffusion of Carbon Atoms In $L1_2$-Fe_3Al Surfaces. A Density Functional Theory Study

Gustavo E. Ramirez-Caballero, Materials Science and Engineering Program, Texas A&M University, 3003 TAMU, Universidad Industrial de Santander, College Station, TX 77843 and Perla B. Balbuena, Chemical Engineering, Texas A&M University, 3122, TAMU, College Station, TX 77843-3122.

Density functional theory is used to study the adsorption, absorption, and diffusion of carbon atoms in $L1_2$-Fe_3Al surfaces. The carbon atoms prefer to locate in certain positions characterized by strong bond energies between carbon and iron atoms. Such strong adsorption determines the formation of carbon structures such as graphite at high concentration of carbon on the surface. The presence of carbon atoms on the surface induces lattice deformations that facilitate the diffusion of carbon inside the surface.

185f Guaranteed Accuracy for Large-Scale Semi-Definite Programs in Electronic Structure Calculations

Frerich J. Keil[1], Denis Chaykin[1], and Christian Jansson[2]. (1) Chemical Engineering, Hamburg University of Technology, Eissendorfer Str. 38, Hamburg, 21073, Germany, (2) Reliable Computing, Hamburg University of Technology, Eissendorfer Str. 38, Hamburg, 21073, Germany

The problem of determining the ground state energy of a system of N electrons, known as the electronic structure problem, is one of the central challenges in quantum chemistry. Traditionally, this problem gives rise to large linear or nonlinear Hermitian eigenvalue problems. But the reduced density matrix (RDM) method allows to solve this problem much faster by using large-scale semidefinite programming ([1], [2]). Semidefinite programms (SDPs) are a class of convex optimization problems which can be summarized as maximization of a linear function on the intersection of a linear affine space and the convex cone of block-diagonal positive semidefinite matrices [3].The full set of sufficient representability conditions for the RDM is not known. A consequence is that the semidefinite program is a relaxation, and thus yields a lower bound of the ground state energy.

In the literature it has been pointed out that the semidefinite programming problems must be solved to high accuracy, typically 7 decimal digits for the optimal value. However, sometimes, owing to the use of floating point arithmetic in combination with ill-conditioning and degeneracy, erroneous results may be obtained. For example, numerical experiences show that for the semidefinite software packages SDPA and SDPARA the accuracy of the optimal value is very poor for some mole-cules such that the lower bound is not below but above the full CI ground state energy.

In our work we present a novel and time efficient algorithm for calculating rigorous lower bounds for the ground state energy of molecules employing the RDM approach; that is, the algorithm provides a guaranteed accuracy. The algorithm is based on the semidefinite programming method described in [4]. By carefully postprocessing the output of a semidefinite programming solver, all rounding errors due to floating point arithmetic are rigorously estimated. The ground state energy of molecules taken from Benchmark problems presented in [1] were calculated by the RDM approach in a semidefinite mathematical form, and then compared with our rigorous results. Additionally, our results were compared with coupled cluster (CCSD) and Hartree-Fock energies. With our algorithms an upper bound of the RDM SDP optimal value can also be computed.

References

[1] Mituhiro Fukuda, Bastiaan J. Braams, Maho Nakata, Michael L. Overton, Jerome K. Percus, Makoto Yamashita, and Zhengji Zhao. Large-scale semidefinite programs in electronic structure calculation. Math. Program., 109(2):553–580, 2007

[2] Zhengji Zhao, Bastiaan J. Braams, Mituhiro Fukuda, Michael L. Overton, and Jerome K. Percus. The

reduced density matrix method for electronic structure calculations and the role of three-index representability conditions. The Journal of Chemical Physics, 120(5):2095–2104, 2004.

[3] Aharon Ben-Tal, Arkadi Nemirowski. Lectures on Modern Convex Optimization.

SIAM, Philadelphia (2001)

[4] C. Jansson, D. Chaykin, and C. Keil. Rigorous Error Bounds for the Optimal Value in Semidefinite Programming. SIAM Journal on Numerical Analysis, 46(1):180–200, 2007.

185g First Principle Simulations of Re3 Metal Cluster Binding on An Alumina Support In Hydrogen Environments

Wenguang Lin, Department of Chemistry and Biochemistry, University of Notre Dame, Notre Dame, IN 46556, Shawn Coleman, Department of Chemical and Biomolecular Engineering, Un iversity of Notre Dame, Notre Dame, 46556, and William Schneider, Department of Chemical and Biomolecular Engineering, University of Notre Dame, 180 Fitzpatrick, Notre Dame, IN 46556.

Metal particles adsorbed on an oxide support are the archetypal heterogeneous catalyst. As interest increases in developing well-defined discrete metal particles on supports, it becomes important to characterize the metal support interaction and in particular its dependence on reaction environment. Gates et al. have demonstrated the preparation and characterization of discrete Re3 clusters on gamma-alumina and have reported that these clusters both are active for hydrogenation catalysis and retain their integrity after reaction(1). In this work we use density functional theory (DFT) simulations to characterize the structure, bonding, and equilibrium composition of discrete Re3 clusters on two limiting models of an á-alumina surface, a hydrogen-free "dry" model and a fully hydroxylated "wet" model. Supercell DFT calculations show that a Re3 cluster binds strongly on the dry surface in registry with three-fold surface oxygen sites, and that this cluster is stable to fragmentation. Binding is driven by charge transfer to undercoordinated surface Al atoms. Molecular adsorption on the wet surface is relatively weak, but spillover of hydrogen from support to cluster greatly enhances bonding. Both results are consistent with Re-surface binding mediated by individual Re-O surface bonds. Based on these results, cluster models are used to probe the ability of supported Re clusters to adsorb hydrogen. An equilibrium coverage of six hydrogens per Re3 cluster are expected under conditions representative of hydrogenation catalysis.

Reference:

(1) Fung, A. S.; Tooley, P. A.; Kelley, M. J.; Koningsberger, D. C.; Gates, B. C. Journal Of Physical Chemistry 1991, 95, 225.

185h First-Principles Theoretical Analysis of Carbon Allotropes and Nanostructures Formed through Exposure of Multi-Walled Carbon Nanotubes to Hydrogen Plasmas

Tejinder Singh[1], Michael J. Behr[2], Andre R. Muniz[1], Eray S. Aydil[2], and Dimitrios Maroudas[1]. (1) Department of Chemical Engineering, University of Massachusetts, 159 Goessmann Laboratory, 686 N Pleasant Street, Amherst, MA 01003, (2) University of Minnesota, 421 Washington Avenue SE, Minneapolis, MN 55455

Carbon forms a variety of different allotropes, such as cubic diamond, hexagonal diamond, graphite, fullerenes, and carbon nanotubes. These carbon allotropes have numerous technological applications. For example, due to its unique properties (such as extreme hardness, high electrical resistivity, high thermal conductivity, and chemical inertness), diamond is used widely in the fabrication of machine tools, optical coatings, X-ray windows, and light-emitting optoelectronic devices. We examine the synthesis of diamond by treatment of carbon nanotubes and carbon nanofibers with H atoms from a H_2 plasma. We observe that upon H_2 plasma exposure of multi-walled carbon nanotubes (MWCNTs) over the temperature range from 300 K to 1073 K, an amorphous carbon matrix is produced with carbon nanocrystals embedded in it.

In this context, we present a detailed theoretical analysis of the various crystalline phases of C observed upon exposing MWCNTs to a H_2 plasma, as well as the potential role of H in determining the structures of the resulting C nanocrystals. In our studies, we use a synergistic combination of high-resolution transmission electron microscopy (HRTEM), electron diffraction, molecular-dynamics (MD) simulations and first-principles density functional theory (DFT) calculations. Our DFT calculations are carried out within the generalized gradient approximation and employ plane-wave basis sets, ultra-soft pseudopotentials, and supercell models. In our MD simulations, H interactions with MWCNTs and the resulting structural relaxations are modeled using the Adaptive Intermolecular Reactive Empirical Bonding Order (AIREBO) potential.

Structural characterization of the nanocrystals formed yields various distinct crystalline carbon phases consistent with a face-centered cubic (fcc) lattice with lattice parameter a = 0.425 nm, a body-centered cubic (bcc) lattice with a = 0.30 nm, a diamond lattice with a = 0.357 nm, and a hexagonal diamond (lonsdaleite) lattice with lattice parameters a = 0.252 nm and c = 0.412 nm. Our theoretical analysis reveals that structural transformations of MWCNTs are triggered initially by the H-induced formation of inter-shell sp^3 C-C bonds. Consequently, a specific alignment of concentric graphene walls results in the formation of inter-shell C-C bonds in the MWCNTs that constitute the seed units, which may nucleate a specific crystalline C phase such as diamond or lonsdaleite (the most abundant ones in our experiments). Furthermore, our results indicate that a new phase of carbon with a body-centered tetragonal lattice with a basis of 4 C atoms (BCT-C4, with lattice parameters a = 0.437 nm and c = 0.252 nm), also referred to as rectangulated carbon, can account for reflections from the electron diffraction that are not matched by the diamond-cubic or lonsdaleite crystalline structures. In addition, we provide other possible interpretations for the experimentally observed crystalline C phases through the incorporation of H atoms, at proper concentrations, in interstitial sites of cubic carbon phases.

185i A Quantum Chemical Approach to Understanding How Gas Phase Conditions Affect the Thermodynamics and Kinetics of Oxidation Catalysis

Rachel B. Getman, Department of Chemical and Biomolecular Engineering, University of Notre Dame, 182 Fitzpatrick Hall, Notre Dame, IN 46556 and William Schneider, Chemical and Biomolecular Engineering, University of Notre Dame, 182 Fitzpatrick, Notre Dame, IN 46556.

One key challenge facing chemical engineers is being able to control the rates of chemical reactions, and therefore much research has focused on understanding catalysis. Heterogeneous catalysis is a popular type and has been implemented into many common processes such as fuel cells, ammonia synthesis, and emissions control. NO oxidation to NO_2 is an example of a heterogeneously catalyzed reaction for the remediation of NO_x from emissions. In this work we use Density Functional Theory (DFT) to examine the thermodynamics and kinetics of NO oxidation catalysis over Pt, the most common catalyst for the reaction. The enthalpy of NO oxidation is quite small, -60 kJ/mol, and therefore, NO_2 formation is best promoted at high pressures, conditions which induce high concentrations of O on the catalyst surface. We show that the surface O concentrations generated under an NO oxidation atmosphere have strong effects on both the thermodynamics and kinetics of different steps in the overall reaction. The results provide insights toward formulating an NO oxidation mechanism and can likely be used to describe other transition metal-catalyzed oxidations as well.

185j Modeling Short and Long-Range Correlations In Polyelectrolyte Brushes

Tao Jiang and Jianzhong Wu. Chemical and Environmental Engineering, University of California, Riverside, Bourns Hall A242, 900 University Avenue, Riverside, CA 92521

The properties of a polyelectrolyte brush depend on a broad range of parameters including the characteristics of the polyions, salt concentration, ion valence, and solvent conditions. We present a non-mean-field theory for polyelectrolyte brushes and discuss the effects of salt concentration, polymer grafting density, and chain length on the polyion configurations, the electrostatic potential, and the distributions of both counterions and coions. Comparison will be attempted with pertinent experimental results and with established theoretical predictions. Special attention will be given to systems containing multivalent counterions wherein the electrostatic interactions are coupled with the excluded-volume effects that defy typical mean-field calculations and scaling analysis. We found that multivalent counterions result in a non-monotonic swelling of the brush that is qualitatively different from that for the case of monovalent counterions. In the limit of low salt concentration, trivalent counterions may self-organize into a one-dimensional wave-like structure in between tethered polyions and lead to a first-order phase transition similar to that for a neutral brush in a poor solvent.

185k Mechanism of Cellulose Hydrolysis : Molecular and Experimental Perspective

Suma Peri, Chemical Engineering, Texas Tech University, Lubbock, TX 79409-3121

Text Not Available.

Using Molecular Modeling to Understand Enzymatic Cellulose Hydrolysis

Xiongce Zhao[1], Everett O'Neal[2], Clare McCabe[2] and Michael E. Himmel[3]

[1]*Center for Nanophase Materials Sciences, Oak Ridge National Laboratory, Oak Ridge, TN 37831*
[2]*Department of Chemical Engineering, Vanderbilt University, Nashville, TN 37235*
[3]*National Renewable Energy Laboratory, Golden, CO 80401*

Cellobiohydrolase I (CBH 1) from *Trichoderma reesei* is one of the most active enzymes in the hydrolyzation of cellulose. CBH 1 is a multi-domain enzyme, consisting of a large catalytic module containing an active site tunnel and a small cellulose binding module joined to one another by a 27 residue linker peptide. As an exoglucanase CBH 1 hydrolyzes cellulose in a "processive" manner; the exact mechanism of is not fully understood. As part of an ongoing effort to better understand the hydrolysis process, molecular dynamics simulations that focus on determining the function of the linker peptide have been performed. Although the sequence of the linker peptide is known and it has been shown that the linker is important for enzymatic activity, the role of the linker is yet to be determined. Results from molecular dynamics simulations in which the motion and conformation of the linker adsorbed on a cellulose surface in an aqueous environment will be presented and our simulation data compared with the hypothesis that the linker behaves like a spring, which helps the enzyme to move along a cellulose chain. Since this spring-like motion is a key element in the processive depolymerization of cellulose by Cel7A, our efforts to understand this mechanism, and thus in the future to engineer more capable cellobiohydrolases, is a key element in making the conversion of biomass to biofuels more efficient.

185m A Computational Study of C_{60}-Pentacene Heterojunction Interfaces

Rebecca A. Cantrell[1], Paulette Clancy[1], Dan Dougherty[2], and Steven W. Robey[2]. (1) School of Chemical and Biomolecular Engineering, Cornell University, Ithaca, NY 14853, (2) Chemical Science and Technology, National Institute of Standards and Technology, Gaithersburg, MD 20899-8300

The morphology of the C_{60}/pentacene heterojunction, of interest for organic photovoltaics applications, is not well understood. On an atomic scale, the supplement of molecular simulation to experimental evidence provides insight into an otherwise enigmatic system. All-atom Molecular Dynamics simulation techniques were used to elucidate the dynamic behavior of small numbers of C_{60} molecules on the surface of crystalline pentacene as a probe of the molecular-level interactions between C_{60} and pentacene. The ultimate single molecule probe of the pentacene surface, a single C_{60} admolecule, exhibited an anisotropic diffusion pattern that lingered in energetically preferred sites in the [1 -1 0] direction that intercepts the 0, ½, 0 point in the unit cell. An Arrhenian analysis of this diffusion data gave estimates for the prefactor, D_0, and energy barrier, E_a , of 2×10^{-3} cm^2/s and 0.1 eV, respectively. Surface diffusion of one C_{60} molecule on pentacene is significantly more rapid (by about 1-2 orders of magnitude) than if even one additional C_{60} admolecule is present, implying that the C_{60}-C_{60} cohesion interaction is stronger than the C_{60}-pentacene adhesion interaction. Simulations with up to four C_{60} molecules, a practical limit of an all-atom approach, reinforced this suggestion that C_{60} likes to dewet a pentacene surface. Collaborators Dan Dougherty and Steve Robey have deposited C_{60} on two layers of flat pentacene on Ag(111), and have observed (via low temperature STM images) nanowires of C_{60} in between pentacene rows at relatively low C_{60} coverages; at higher C_{60} coverages, they observe a combination of nanowires in between pentacene rows and disordered C_{60} thin films. While the systems considered in experiment are not exactly the same as those in simulation, both suggest that both the C_{60}-pentacene adhesion and the C_{60}-C_{60} cohesion influence the C_{60} thin film structure. The question of what exactly determines the relative order of the C_{60} thin films on pentacene is yet to be answered.

185n A Monte Carlo Simulation Study of Strain-Relief Patterns In Ag / Pt (111) Heteroepitaxy

Joshua D. Howe, Dept. of Chemical Engineering, The Pennsylvania State University, University Park, PA 16802 and Kristen A. Fichthorn, Dept. of Chemical Engineering, The Pennsylvania State University, University Park, PA 16802.

Experimental studies of Ag/Pt(111) heteroepitaxy show rich structures that arise due to strain at the Ag/Pt interfaces [1]. Strain-relief structures, such as stripes and patterns consisting of hexagons and triangles, occur in the 2nd monolayer of Ag on Pt(111). In our work, we investigate the origins of these patterns. The Ag(111) surface has a Shockley-type surface state that is sensitive to strain [2]. In first-principles studies with density functional theory (DFT) the pair interactions between Ag atoms have been characterized up to the 53rd neighbor and are shown to correlate with a one-dimensional theoretical expression for the interaction as a function of separation [2]. However, the DFT result exhibits angular variation in the interaction and takes on a two-dimensional character that is not captured by the theory. Here, we show that this angular variation is significant. We use Monte Carlo (MC) simulations to study the structure of Ag layers in which the atoms have pair interactions described by both DFT and theory. We show that, with the DFT interactions, we can reproduce all the key aspects of the striped patterns seen experimentally, including the crystallographic orientation of the stripes and the alternation of the stripes between fcc and hcp registry. These results suggest that electronic surface states play a significant role in determining surface structures and that the angular variations in surface-state-mediated interactions are significant in determining system behavior.

[1] H. Brune, H. Röder, C. Boragno, K. Kern, Phys. Rev. B, 49 2997 (1994).

[2] W. Luo, K. Fichthorn, Phys. Rev. B, 72 115433 (2005).

185o Lattice Kinetic Monte Carlo Simulations of Continuous Systems

Xiao Liu[1], Warren D. Seider[2], and Talid Sinno[2]. (1) Department of Chemical and Biomolecular Engineering, University of Pennsylvania, 220 S. 33rd Street, 311-A Towne Building, Philadelphia, PA 19104, (2) Chemical and Biomolecular Engineering, University of Pennsylvania, Philadelphia, PA 19104

The lattice kinetic Monte Carlo method (LKMC) is an efficient approach for simulating dynamical evolution in microscopic systems, because it coarse-grains out the details of atomic vibration that severely limits the scope of molecular dynamics (MD) simulations while retaining atomistic resolution. However, the confinement of atomic positions to a rigid lattice necessarily reduces the available configurational degrees of freedom in a system. This constraint can become very important at elevated temperatures.

In previous work, a bond-counting LKMC model for vacancy diffusion and aggregation in silicon was generated. It was shown that the lattice constraint led to large errors in the predicted dynamics of vacancy clustering at high temperature because off-lattice configurations could not be explicitly captured in the rigid lattice model [1]. The missing configurational degrees of freedom were implicitly accounted for by regressing the bond energies to data generated by MD simulations [2]. While this approach successfully captured important high temperature entropic contributions, the lack of transferability to different temperatures limits its application to realistic processes such as crystal growth.

In this work, we present a LKMC framework to simulate continuous systems. Previous Metropolis Monte Carlo modeling studies have shown that lattice representations are able to correctly capture the phase behavior of continuous fluids if the discretization is sufficiently fine [3]. Here, we investigate the effect of lattice spacing on the ability of a LKMC simulation to capture the correct dynamical behavior by making detailed comparisons to molecular dynamics simulations. We discuss the calculation of rate barriers from a prescribed interatomic potential such as the Lennard-Jones model. We show that it is possible to generate the correct dynamics with LKMC for sufficiently discretized grids.

[1] S. S. Kapur, M. Prasad, J. C. Crocker, T. Sinno, Role of configurational entropy in the thermo-dynamics of clusters of point defects in crystalline solids, Phys. Rev. B 72, 014119 (2005).

[2] J. Dai, W. D. Seider and T. Sinno, Lattice kinetic Monte Carlo simulations of defect evolution in crystals at elevated temperature. Molecular Simulation, 32, 305 (2006).

[3] A. Z. Panagiotopoulos, On the equivalence of continuum and lattice models for fluids. J. Chem. Phys., 112, 7132 (2000).

185p Molecular Dynamics Simulation of Bhet: 1,4-Benzenedicarboxylic Acid 1,4-Bis(2-hydroxyethyl)Ester

Qifei Wang, Department of Chemical and Bimolecular Engineering,, University of Tennessee, Knoxville, Knoxville, TN 37996-2200, Knoxville, TN TN 37996-220 and David Keffer, Chemical and Biomolecular Engineering, The University of Tennessee, 419 Dougherty Hall, Knoxville, TN 37996-2200.

ABSTRACT:

In this poster, we report on our molecular dynamics (MD) simulation work of 1,4-benzenedicarboxylic acid 1,4-bis(2-hydroxyethyl)ester (BHET). BHET is the monomer associated with the production of polyethylene terephthalate (PET). We perform MD simulations in the NVT and NpT ensembles, using a model in which all atoms are explicitly represented expect for the hydrogen connected to carbons. The hydrogen atoms bound to oxygen are explicitly described to more accurately account for hydrogen-bonding in the system.

We report the following thermodynamic and transport properties as a function of temperature and pressure: (i) fluid density, (ii) heat capacity, (iii) shear viscosity, (iv) self-diffusivity, and (v) thermal conductivity. We examine the intramolecular and intermolecular structure of the fluid through pair correlation functions, with a special examination of the hydrogen-bond network.

We also report our progress on simulation of systems of oligomers of BHET (n = 1 to 10).

185q A Computationally Efficient Algorithm for Accurate Local Energy Minimization of Crystal Structures Containing Flexible Molecules

Andrei V. Kazantsev[1], **Panagiotis G. Karamertzanis**[1], Claire S. Adjiman[1], and Constantinos C. Pantelides[2]. (1) Department of Chemical Engineering, Imperial College London, Centre for Process Systems Engineering, London, SW7 2AZ, United Kingdom, (2) Imperial College London / Process Systems Enterprise Ltd, Centre for Process Systems Engineering, London, SW7 2AZ, UK

Crystal structure prediction of organic molecules from their atomic connectivity diagram has been a key problem in computational chemistry for over a decade. Following recent advances, there exist reliable search methodologies that manage to identify most low-energy minima (including the experimentally determined polymorphs) for crystals of both rigid[1] and flexible[2] molecules. However, a major problem with these techniques is that, in order to be able to search efficiently what is a very large space of possible crystal structures, they have to rely on approximate estimates for the inter-molecular and intra-molecular energy contributions:

µ Inter-molecular electrostatic contributions are represented using atomic (and sometimes other) charges derived from quantum mechanical calculations of the molecule's electrostatic field. Albeit computationally efficient, such representations are not as accurate as those based on multipole moments.

µ In the case of flexible molecules, intra-molecular energy is computed by interpolating the results of quantum mechanical calculations over pre-computed grids of the molecule's flexible torsion angles[2]. The atomic charges may also be expressed as functions of these torsion angles in a similar fashion. Although the interpolants themselves are quite accurate, the distinction between "flexible" and "rigid" degrees of freedom in the molecule introduces a degree of approximation which miscalculates the actual positions of the atoms in the molecule within the crystal lattice, and therefore the corresponding inter-molecular interactions.

Because of the above reasons, the absolute energy and, consequently, the ranking of low-energy crystal structures identified by state-of-the-art search methods are not reliable. They, therefore, have to be corrected by re-minimization of the crystal lattice energy using more accurate models. One of the most advanced algorithms available for this purpose is DMAFlex[3]. This incorporates a two-level optimization procedure:

µ The outer optimization manipulates the values of the flexible degrees of freedfom, q^F, (such as torsion angles, bond angles, etc.) and performs a quantum mechanical calculation to determine the minimum energy molecular conformation corresponding to the current set of values.

µ The inner optimization identifies the crystal structure that minimizes the inter-molecular energy based on the molecular conformation currently fixed by the outer optimization. The electrostatic contributions to the inter-molecular energy are derived via an anisotropic distributed multipole model (up to hexadecapole level) computed[4] from the results of the quantum mechanical calculation at the outer level.

The above approach is quite accurate as demonstrated by its applications to several systems of interest[5]. However, the incorporation of a complete quantum mechanical calculation within an optimization loop results in high computational cost. Moreover, the derivation of a new set of multipoles after each conformational change calculation is also computationally demanding. In addition, multipole expansion of the isolated molecule charge density may result in non-differentiabilities. Furthermore, the use of a gradient-free (simplex) algorithm for the outer optimization, which can overcome small discontinuities, limits the extent of molecular flexibility that can be handled with a reasonable number of outer iterations.

This paper presents a novel energy minimization algorithm that is significantly more computationally efficient than the one presented above, whilst being of equivalent accuracy. The algorithm also employs a two-level optimization scheme but without the need for a quantum mechanical calculation at every outer iteration:

µ Intra-molecular energies are estimated using a local approximate model based on a quadratic Taylor expansion of intra-molecular energy in terms of the flexible molecular degrees of freedom, q^F. The expansion is constructed around a base point derived from the intra-molecular energy and its first and second derivatives computed via an isolated-molecule wavefunction calculation.

µ The above wavefunction calculation is also used to compute a distributed multipole expansion of the isolated-molecule charge density, where moments up to hexadecapole are expressed with respect to the local axis system of each atom. The multipoles are then used to model the conformationally dependent intermolecular electrostatic interactions by assuming that they remain invariant with respect to the local axis system.

µ The above define local approximate models which are valid within a certain distance (in q^F space) of the point at which they were derived. The frequency with which these local models are updated is automatically and independently adjusted by monitoring their molecule-dependent sensitivity to molecular conformation as the minimization progresses.

µ A quasi-Newton minimization algorithm is used for both the inner and outer optimization problems. This ensures rapid convergence even when there are many flexible degrees of freedom.

The validity of the proposed methodology has been tested by minimising an extensive set of experimentally determined and computer generated crystal structures containing flexible molecules. The results obtained are essentially identical (within numerical error) with those of the earlier scheme, but obtained at a fraction of the computational cost. In practice, we have found that many optimizations can be done using a single quantum mechanical calculation carried out at the initial point.

As a result, the proposed methodology addresses the limitations identified in previous algorithms. It also opens some interesting possibilities for future work, e.g. in terms of removing the distinction between "flexible" and "rigid" degrees of freedom, and also, potentially, incorporating this more rigorous energy calculations directly within the global search techniques.

Reference List

[1] Karamertzanis, P. G. and Pantelides, C. C. *J. Comput. Chem.* **2005**, *26*, 304-324.

[2] Karamertzanis, P. G. and Pantelides, C. C. *Mol. Phys.* **2007**, *105*, 273-291.

[3] Karamertzanis, P. G. and Price S. L.; *J. Chem. Theory Comput.* 2006, 2, 1184.

[4] Stone, A. J. *J.Chem.Theory Comp.* **2005**, *1*, 1128-1132.

[5] Karamertzanis *et al.*; *J. Comput. Chem.* **2007**, B, Vol. 111, 19, 5326-5336

185r Self-Consistent Thermostatting of Molecular Dynamics Simulation In the Presence of Inhomogeneous, Non-Equilibrium Fields through Multiscale Modeling

Ruichang Xiong, Rebecca L. Empting, Ian C. Morris, Jared T. Fern, and David J. Keffer. Chemical Engineering, University of Tennessee, 419 Dougherty Hall, Knoxville, TN 37996-2200

Spatially inhomogeneous non-equilibrium fields can give rise to non-uniform generation of heat. Molecular dynamics simulations performed in the presence of inhomogeneous, non-equilibrium fields therefore require thermostatting capable of adding or removing energy with spatial selectivity. If the field is not isotropic, it is also necessary to decouple the thermostats for each dimension of the system. In this work, we demonstrate that a thermostatting scheme with both spatial and dimensional selectivity is capable of generating a uniform temperature profile in the presence of an anisotropic and spatially inhomogeneous non-equilibrium field. We show that the resulting density profile is consistent with both a molecular-level theoretical prediction from statistical mechanics as well as the solution of the continuum-level set of differential equations describing the conservation of mass and momentum. Finally, we show that for fluids with non-negligible intermolecular interactions, the continuum-level description requires a mechanical equation of state generated by the molecular dynamics simulation in order to provide consistent results.

185s Using Molecular Simulation to Study Wetting Behavior

Eric M. Grzelak, Dept. of Chemical and Biological Engineering, University at Buffalo, 303 Furnas Hall, Buffalo, NY 14260 and Jeffrey R. Errington, Chemical and Biological Engineering, University at Buffalo, 303 Furnas Hall, Buffalo, NY 14260.

The study of wetting properties at the micro and nanoscopic level is well suited to computational techniques due to its small length scale. The characteristics of wetting in these diminutive systems are exploited in fields ranging from microfluidics to labs-on-a-chip. With this presentation we outline our recent computational effort to expand the understanding and predictability of how fluids wet rough surfaces with length scales O(1) nanometers and below. Using grand canonical transition matrix Monte Carlo simulations we determine surface free energies which lead to spreading coefficients and prewetting saturation points for systems below and above the wetting temperature respectively. Surfaces investigated are comprised of static particles in crystalline configurations sculpted to differing degrees of roughness. Quantitative techniques measure the roughness and then compare data to models of wetting devised by Wenzel and Cassie. Both spreading coefficients and prewetting saturation data are extrapolated to better determine wetting temperatures, which are then also compared to the differing degrees of roughness.

185t Comparison of Functionalized Amine Energetics for CO_2 Capture

Elaine Mindrup and William F. Schneider. Department of Chemical and Biomolecular Engineering, University of Notre Dame, 182 Fitzpatrick Hall, Notre Dame, IN 46556

Growing concern about global climate change is driving the demand for efficient, cost-effective, and durable processes to separate CO_2 from the flue gas of coal-fired power plants. The most common approach to CO_2 separation is chemical absorption using aqueous ethanolamine. In this reaction, two amine molecules combine with one CO_2 to make a carbamate anion and ammonium cation. In this work, we use first principles simulations to examine the dependence of the thermodynamics of the carbamate reaction on the amine. We calculate gas-phase and solution-phase energetics over the full molecular conformational space, analyze the contributions of carbamate and ammonium formation to overall reaction energies, and discuss the potential to tune reactivity by modification of functional groups. We further consider the implications of these results for the selection of amine functional groups for reactive ionic liquids (or so-called "task specific" ionic liquids) under study as alternatives to the ethanolamine system. Ionic liquids present a unique reaction environment for formation of carbamate, and we consider the consequences of cation vs. anion tethering of amines.

185u Reverse Engineering the Human Platelet: A Computational Framework for Predicting Platelet Activation

Jeremy E. Purvis[1], Manash S Chatterjee[2], Lawrence F Brass[1], and Scott L. Diamond[2]. (1) Medicine, University of Pennsylvania, 5140 Vagelos Research Laboratories, 3340 Smith Walk, Philadelphia, PA 19104, (2) Chemical and Biomolecular Engineering, University of Pennsylvania, 5140 Vagelos Research Laboratories, 3340 Smith Walk, Philadelphia, PA 19104

Platelets respond to endothelial injury or activating agonists through a complex sequence of signaling mechanisms. To understand how these events are coordinated in time and space, we developed a computational model of the human platelet by combining existing kinetic information, electrochemical calculations, measurements of platelet ultrastructure, novel experimental results, and published single-cell data. The model accurately predicted (i) steady-state concentrations for intracellular calcium, inositol 1,4,5-trisphosphate, diacylglycerol, phosphatidic acid, phosphatidylinositol, phosphatidylinositol phosphate, and phosphatidylinositol 4,5-bisphosphate, (ii) transient increases in intracellular calcium, inositol 1,4,5-trisphosphate, and Gq•GTP in response to ADP, and (iii) the volume of the platelet dense tubular system. A more stringent test of the model involved stochastic simulation of individual platelets, which are known to display an asynchronous calcium spiking behavior in response to ADP. The model reproduced the measured frequency distribution of spiking events and demonstrated that asynchronous spiking was a consequence of stochastic fluctuations due to the small volume of the platelet. These simulations offer new insight into platelet enzymatic regulation, receptor signaling, and the physical structure of the cell. As such, the model provides a novel approach for elucidating normal platelet signaling behavior as well as designing pharmacological strategies for platelet therapy.

Web Page: bloodsystemsbiology.org

185v Atomistic Simulation of Membrane Puncture Using a Carbon Nanotube Syringe

Vamshi K. Gangupomu, Chemical and Biological Engineering, Drexel University, 3141 Chestnut Street, Drexel University, Philadelphia, PA 19104 and Franco M. Capaldi, Mechanical Engineering and Mechanics, Drexel University, 3141 Chestnut Street, Drexel University, Philadelphia, PA 19104.

The advent of Nanomedicine has stirred increased interest in the interactions of nanomaterials or nanoparticles with biological molecules. Devices and structures in the 100 nm scale may revolutionize the way drugs are delivered into cells and even organelles. Nanomaterials such as fullerenes, carbon nanotubes and metallic nano particles are being studied for possible applications in development of these devices. The most promising of these nanomaterials are the carbon nanotubes (CNTs) due to their unique properties and dimensions. Suggested applications for CNTs range from medical and biological applications such as nanosyringes, force probes, drug delivery agents and gene therapy.

Nanosyringes are devices that could be used to transport therapeutic molecules or genetic information across the cell membrane. Existing technologies to deliver molecules into a cell include lipid fusion, viral vectors, gene guns, and microinjectors. Of all the above mentioned methods, microinjection being a physical method is easy to use and control. However, microinjection is presently performed using glass micropipettes. The glass capillaries are either too big and could cause difficulties such as membrane rupture and damage crucial cell organelles. Physical disruption of the cell plasma membrane by the micropipette depends upon the applied force magnitude and rate, and the size of the pore created by the tip. The dimensions of carbon nanotubes, their high aspect ratio and mechanical robustness make them suitable for their use as micropipette tips.

Carbon nanotubes were also suggested as scanning probe microscopy tips because of their precise definition and robustness. CNTs have a very high Young's modulus (~ 1TPa) and tensile strength and their small diameters and shape increase the accuracy of the probe and are effective in measuring sharp recess, high aspect ratio samples.

A thorough understanding of the interactions between nanoinjector and the cell plasma membrane is necessary. The dimensions of the membrane and the weak forces involved makes it a challenge in handling and manipulating cells and the atomic level interactions between the membrane and CNTs play a very important role. We use Molecular Dynamics simulations to study the interactions of penetrating nanotubes with a lipid bilayer. The minute forces involved in these interactions make them difficult to interpret from experimental results. Atomistic simulations provide excellent details of the physical interactions and an accurate measurement of the forces involved.

Preliminary simulations involving a simple bilayer system of phosphotidylcholine (POPC), CNT and water have been performed. The interactions between the lipid bilayer and the nanotube during the insertion and withdrawal of the CNT, as in a nanosyringe, were studied. These simulations suggested a quick healing of the cell membrane after rupture during the insertion. Also the forces involved during the insertion and withdrawal were studied, with the nanotube acting like a force probe tip. These forces showed a significant decrease in the membrane rupture force compared to the glass capillaries and are in agreement with previously observed rupture forces using CNT tips. Additional simulations are being performed to study the effects of variables such as insertion velocity, orientation and membrane composition on the forces involved and the healing process.

185w Multiscale Monte Carlo Study of Epidermal Growth Factor Receptor Diffusion and Dimerization

Stuart D. Collins, Chemical Engineering, University of Delaware, 150 Academy St, Newark, DE 19716, Abhijit Chatterjee, Theoretical Division, Los Alamos National Laboratory, T-12 MS B268, Los Alamos, NM 87545, and Dion Vlachos, Director of Center for Catalytic Science and Technology (CCST), University of Delaware, Newark, DE 19716.

The ErbB family of receptors is dysregulated in a number of cancers, and the signaling pathway of this receptor family is a critical target for several anti-cancer drugs. Therefore, a detailed understanding of the mechanisms of receptors activation is critical. However, despite a plethora of biochemical studies and single particle tracking experiments, the early molecular mechanisms involving epidermal growth factor (EGF) binding and EGF receptor (EGFR) dimerization are not as well understood. Due to the large disparity of time and length scales involved in receptor dimerization reactions, we adapt the coarse-grained Monte Carlo (CGMC) simulation framework to enable the simulation of in vivo receptor diffusion and dimerization.

On-lattice kinetic Monte Carlo (KMC) simulations have extensively been applied to numerous systems. However, their applicability is severely limited to relatively short time and length scales. Recently, the CGMC method was introduced to greatly expand the reach of the lattice KMC technique. Herein, we extend the previous spatial CGMC methods to multicomponent species and/or site types. Numerical examples are presented to demonstrate the method. Furthermore, we introduce the concept of homogenization at the stochastic level over all site types of a spatially coarse grained cell. Homogenization provides a novel coarsening of the number of processes, an important aspect for complex problems plagued by numerous microscopic processes (combinatorial complexity). As expected, the homogenized CGMC method outperforms the traditional KMC method on computational cost while retaining good accuracy.

Using the CGMC method, spatial modeling of ligand-mediated membrane receptor dimerization reaction dynamics was peformed. Furthermore, the simulations demonstrate the importance of spatial heterogeneity in membrane receptor localization. Mathematical models, especially one that takes into account spatial heterogeneity, show mechanistic understanding of receptor activation that may in turn enable improved future cancer treatments.

185x Investigating Different Levels of Coarse-Graining in Models for Water

Kevin R. Hadley and Clare McCabe. Department of Chemical and Biomolecular Engineering, Vanderbilt University, Box 1604 Station B, Nashville, TN 37235

Coarse-grained (CG) simulations are widely used for observing molecular phenomena, such as self-assembly in biological systems. CG models are typically developed by optimizing the model's potentials to match properties obtained from atomistic simulations. Typically in simulations of self-assembly an explicit water solvent is used; however, for coarse-graining methods that require an explicit mapping of atomistic properties to the CG level for potential optimization, water is usually mapped with one bead representing one water molecule, forcing the solvent to be the most computationally expensive component of the simulation. Although other methods where a bead in the CG model represents multiple water molecules are more computationally efficient, the derivation of their potentials are not systematic and may not be robust. To address this problem we have devised a novel strategy to coarse-grain water to a multiple molecule per bead model. Results will be presented for several CG water models solvating hydrophobic and hydrophilic CG molecules to demonstrate what degree of detail (with respect to water) is necessary to mimic behavior seen in atomistic simulations.

185y Divalent Cations-Dimethylphosphate Complex Formation: Providing Insights into Membrane Fusion In Cells

Zeena K. Issa, Charles W. Manke, Bhanu P. Jena, and Jeffrey J. Potoff. Chemical Engineering, Wayne State University, 5050 Anthony-Wayne Dr., Detroit, MI 48202

Prior x-ray diffraction, light scattering, photon correlation spectroscopy, and atomic force microscopy experiments suggest that SNARE-induced membrane fusion in cells proceeds as a result of calcium bridging opposing bilayers, which leads to the release of water from hydrated Ca^{2+} ions as well as the loosely coordinated water at PO-lipid head groups [1]. It is hypothesized that the combined effect of local dehydration of phosphate head groups, as a result of Ca^{2+} bridging, leads to a destabilization of lipid bilayers and resulting membrane fusion. This hypothesis was tested in the current study by performing atomistic molecular dynamic simulations in the isobaric-isothermal ensemble on dimethylphosphate anions (DMP^-) and Ca^{2+} in water at 298 K and 1.01 bar. Examination of molecular configurations obtained from the molecular dynamics simulations reveals the establishment of self-assembled Ca^{2+} - DMP^- complexes. Calcium ions can bridge two opposing DMP^- molecules, resulting in the formation of a somewhat ring-like structure [2].

Experimental evidence shows that other divalent cations, such as magnesium and barium, may also initiate fusion, but at slower rate [3,4]. In this study, molecular dynamics simulations were preformed to test if other divalent cations self assemble with DMP^- under same molecular dynamics conditions. Simulations show significantly fewer incidents of DMP-Mg-DMP bridges forming. Simulations of Ba^{2+} show no incidents of bridging dimethylphosphate oxygens. Radial distribution functions and number integrals show the depletion of water around DMP^- head groups and hydrated divalent cations upon bridging incidents.

Ab initio calculations were preformed to study the stability of DMP^- and its complexes with divalent cations in the presence of implicit solvent. Calculations were preformed at the B3LYP/6-31G(d,p) level of theory and CPCM solvent model. The data indicate that calcium posses a higher binding affinity to DMP^- than Mg^{2+} or Ba^{2+}. Furthermore, calcium favors participating in ring complexes over forming a single bridge between two DMP^- molecules.

[1] Jeremic, A., Cho, W-J., Jena, B.; J. Biol. Phys. & Chem. 4:139-142 (2004)

[2] Potoff, J., Issa, Z., Manke, C., Jena, B., Cell Biol. Int. 00:00-00 (2008)

[3] Papahadiopoulos, D., Nir, S., Duzgunes, N.; J. Bioenergetics 22:157 (1990)

[4] Ohki, S.; Biochem et Biophys. 1:689 (1982)

185z Investigating Metabolite Turnover Rates Using Constraint-Based Models of Metabolism

Laurence Yang, Radhakrishnan Mahadevan, and William R. Cluett. Chemical Engineering and Applied Chemistry, University of Toronto, 200 College St., Toronto, ON M5S 3E5, Canada

The past few years has seen a significant advancement in the predictive capability of constraint-based models of metabolism due to incorporation of thermodynamic constraints. These models allow us to broaden the scope of prediction and analysis to both reaction fluxes and metabolite concentrations. Taken together, these process variables can now provide insight into additional characteristics of metabolism, such as the turnover rate of each metabolite. Furthermore, these *in silico* investigations can be refined using increasingly available multi-omics data. Using a thermodynamically constrained metabolic model of *Escherichia coli* with additional constraints from multi-omics measurements in the literature, we found that the metabolite turnover rates were bimodally distributed. We also found a correlation between a metabolite's turnover rate and its connectivity. Furthermore, we investigated the degree to which perturbations in each turnover rate affected the rest of the network. For this, we distinguished between perturbations affecting turnover rate values--but not necessarily their distribution--with those affecting the distribution of turnover rates across the entire network. To account for the variability in fluxes and metabolite concentrations in the above analyses, we formulated a fractional program that could be used to define the maximum and minimum ranges of metabolite turnover rates. Based on the fractional programming literature, we reformulated this problem into a mixed-integer linear program that could be efficiently solved for genome-scale metabolic network models. In addition to offering further insights into network properties, we show how metabolite turnover rates have important implications when designing strains for metabolic engineering applications.

Thyroid Receptors (TRs) belong to the family of the Nuclear Hormone Receptors. They are essential in regulation of growth, metabolic rate, cholesterol level and heart rate. In mammals four isoforms are present; TR alpha-1, TR alpha-2, TR beta-1 as well as TR beta-2. In this work we present the process of designing novel Maltose binding protein-Intein- Thymidylate Synthase:: Thyroid Receptor alpha-1 (pMIT::TR alpha-1) biosensor to detect potential TR alpha-1 agonists and antagonists. Resent results of computational analysis of binding modes of the compounds to the TR alpha-1 active side are also shown. In this study Shape Signatures method allowed for fast identification of the compounds which are similar to known TR antagonists and agonist such as NH3, T3 and Triac.

185ab Identify Reaction Pathways Via Constrained Optimization

Jhih-Wei Chu and Kevin R. Haas. Chemical Engineering, University of California, Berkeley, 101A Gilman, UC Berkeley, Berkeley, CA 94720

Methods based on the chain-of-states approach are very useful in bridging timescale gaps for the studies of rare events. However, the requirements of ensuring equal distances between replicas along a chain significantly affect the optimization of objective functions and lead to slow convergence and numerical instability during path optimization. In this work, we present a new algorithm that represents the equal-distance requirements as holonomic constraints. This scheme thus decouples path optimization from equal-spacing replicas under the framework of constrained optimization using Lagrange multipliers. Straightforward implementation of super-linear minimization schemes such as the Adopted Basis Newton Raphson (ABNR) method can thus be achieved. As a result, the quality of convergence and stability of path optimization is greatly improved compared to conventional methods such as the nudged elastic band (NEB) method. In addition to minimum energy paths (MEP), we also apply this method to find the minimum free energy paths (MFEP) in molecular systems. Moreover, a novel hyper plane constraint is also developed to compute the potential of mean force along an optimized path. This set of new methods is applied to study the isomerization of alanine dipeptide and the helix-to-hairpin transition of an amyloid beta peptide. To model the helix-to-hairpin structural transition, as high as 513 replicas were used, resulting in more than 60,000 atoms in the chain for path optimization. This example thus demonstrates the robustness and efficiency of our scheme of constrained path optimization. Finally, outlook of applying the methods developed in this work to simulate transitions in complicated molecular systems will be given.

185ac Role of Intersubunit Hydrogen Bond to Urate Oxidase Stability: Molecular Simulation and Experimental Validation

Zhixia Liu, Diannan Lu, and Zheng Liu. Department of Chemical Engineering, Tsinghua University, Beijing, China

Urate oxidase (UOX, EC 1.7.3.3), a tetrameric "Tunnelling-fold" enzyme for the conversion of uric acid, is effective for the treatment of gout and hyperuricaemia associated with tumor lysis syndrome. The inherent poor stability of UOX to temperature, proteolysis and acidic environment is known to limit its efficacy while the deactivation mechanism is poorly understood, particularly at molecular level.

This presentation started with a molecular dynamic simulation of aqueous UOX, which illustrated that the intersubunit hydrogen bond was essential for maintaining the native conformation of UOX. Also, we have proved the stability improvement of UOX can be achieved by superseding water molecules surrounding UOX surface with solvents of weak compete for the intersubunit H-bond.

Experimental validation was performed using SEC, circular dichroism and florescence spectroscopy to monitor the conformational changes of UOX. It was shown that the secondary and tertiary structure unfolding instead of quaternary structure dissociation occurred during inactivation process of UOX. This demonstrated the crucial role of H-bonding at subunit-subunit interface, as illustrated by the molecular simulation. Furthermore, as predicted by the in silico experiment described above, the presence of methanol and DMSO in aqueous UOX enhanced the stability of UOX and stabilized its tertiary structure as proven by activity assay and fluorescence spectrum results, which also supported the stabilization mechanism by MD simulation.

Chemical modification of UOX was then attempted via two-step aqueous in situ polymerization and fabricated UOX nanogel with a diameter of 20-40nm. Compared to its native counterpart, UOX nanogel exhibited a significantly enhanced stability against low pH, high temperature, and more importantly, proteolysis, proved by activity assay, structural change and integrity. The proposed method is helpful for the further development of UOX formulation in drug delivery system.

451a Programmable Self-Assembly of DNA-Tethered Vesicle Superstructures

Paul Beales, Princeton University, Princeton, NJ 08544 and **T. Kyle Vanderlick**, Yale University, New Haven, CT 06520.

Nature employs lipid structures as a smart, environment-sensitive packaging material that also acts as a two dimensional solvent for proteins, receptors and channels to interact at the interface between these compartments. Even simple vesicles reconstituted in vitro with minimal lipid components can exhibit a zoology of morphological changes, phase behavior and biomimetic properties, for instance budding and fusion. These attributes make liposomes a desirable material for engineering soft, functional containers for confinement and transport of chemicals in novel technologies such as microfluidic platforms or targeted delivery systems. Another bioinspired strategy with technological promise is the programmable assembly of nanoscale and microscale components by the binding of complementary DNA sequences. The digitally-encoded recognition between oligomeric, single DNA strands is highly sensitive to single base mismatches and environmental conditions, affording the potential of fine control over self-assembling structures. Single stranded DNA modified with a hydrophobic group anchors the oligonucleotide to the outer monolayer of fluid vesicle membranes. We demonstrate the assembly of large (100nm) and giant (>5μm) vesicles decorated with complementary DNA sequences. Three regimes of aggregation behavior are observed: no aggregation, stable aggregates and aggregates that grow continually with time. The aggregate size and growth kinetics can be controlled by choice of system parameters, for example the number of DNA per vesicle, the ionic strength of the solution and the electrical potential of the vesicle membrane. The vesicle structures that form are reversible: repeatedly heating and cooling through the DNA melting transition unbinds and rebinds the vesicle assemblies. The melting temperature of DNA is sensitive to the lipid composition of the vesicles. This can be understood by modeling how the inter-membrane repulsive pressure tilts the free energy landscape of the DNA. Phase separation in the membrane can be utilized to confine the DNA to localized regions on the vesicle surface. We find that the partitioning of the DNA between lipid phases is sensitive to the anchoring technique that is employed. These results illustrate that careful engineering of the lipid composition as well as the attached nucleic acid sequences is a valuable tool in controlling the self-assembly of hierarchical superstructures of DNA-tethered vesicles.

451b Controlling the Self-Assembly of Wormlike Micellar Networks by Addition of Colloidal Particles

Matthew H. Helgeson[1], Florian Nettesheim[1], **Eric Kaler**[2], and Norman Wagner[1]. (1) University of Delaware, Newark, DE 19711, (2) Stony Brook University, Stony Brook, NY 11794

Wormlike micelles (WLMs) are a class of self-assembled structures with useful rheological properties. These properties can be controlled by addition of colloidal particles. Specifically, particle addition leads to formation of viscoelastic networks in dilute solutions and enhances the viscoelasticity of semi-dilute, entangled WLM solutions. Particle addition also suppresses shear banding near the isotropic-nematic phase transition. Measurements show that this additional structuring occurs by the direct association of micelles at particle surface. These micelle-particle aggregates increase the compressibility of the micellar fluid, contribute increased viscoelasticity to the WLM network through the contribution of additional network junctions, and suppress shear banding by restricting micellar flow-alignment near the particle surface. Thus, particle addition to WLMs enables the properties of viscoelastic micellar networks to be tuned.

451c Silica-Surface Cleaning with Aqueous Surfactant-Stabilized Gas/Solids Suspensions

V. Andreev, John Prausnitz, and **Clayton Radke**. University of California, Berkeley, CA 94720-1462

We present a new wet-cleaning method to remove strongly adhered 90-nm, Si3N4 contaminant test particles from a smooth silica surface. The proposed cleaning solution consists of a foamed aqueous dispersion of solids. Contaminant removal increases with solids concentration, foam quality, process time, and shear rate and can approach 100 %.

Particle-removal experiments in a simple dip cell indicate that dispersed solids are necessary for contaminant removal. Exponential decline of surface concentration of adhered particles with number of immersion/withdrawal events and a linear increase in the rate of particle removal with solids concentration indicate that the removal mechanism is binary collision between Si3N4 contaminant particles and the solids. Foam or surfactant solution or aqueous solvent alone are not effective in particle removal.

A binary-collision rate model is derived to predict particle detachment. Comparison of the rate model with experimental removal data is good. Three adjustable parameters arise when no foam is employed: two first-order rate constants and the fraction of particles with each rate constant. Prediction of particle removal with a foamed-solids dispersion requires an additional influence parameter. The new binary-collision rate law provides a powerful design tool to assess the role of cleaning parameters including process time, shear rate, dispensing flow rate, system geometry, solids concentration, and foam quality.

451d Molecular Dynamics Simulations of Surfactant and Lipid Assemblies

Ronald Larson, Zuwei Wang, Barry Shang, Hwankyu Lee, and Senthil Kandasamy. University of Michigan, Ann Arbor, MI 48109

We use atomistic and coarse-grain molecular dynamics simulations to study the interactions of surfactant micelles and lipid monolayers and bilayers with polymers and proteins. Specific problems include atomistic molecular dynamics simulations of an anionic sodium dodecyl sulfate (SDS) micelle and a nonionic poly(ethylene oxide) (PEO) polymer in aqueous solution. We show that the polymer resides on the micelle surface and at the hydrocarbon-water interface, leading to a selective reduction in the hydrophobic contribution to the solvent-accessible surface area of the micelle. The association is mainly driven by hydrophobic interactions between the polymer and surfactant tails, and produces the commonly accepted "beaded necklace" structure of the SDS-PEO complex. We also investigate "hydrophobic mismatch" in the length of hydrophobic domains of transmembrane peptides relative to the width of the hydrophobic domain of bilayers using model "KALP" peptides consisting of alanine and leucine lipophilic residues capped at both ends by lysines. We find that the peptide/bilayer system compensates for mismatch through a combination of peptide tilting, local bilayer bending, and lysine side-chain "snorkeling," in the case that the hydrophobic domain of the peptide is shorter than that of the lipid bilayer. Finally, the conditions under which cationic dendrimers can induce pore formation in bilayers are studied by coarse-grained MD simulations and the results compared favorably with recent experiments.

451e Elucidating Chiral Discrimination Mechanisms In Polysaccharide-Based Sorbents Using Experimental Probing and Molecular Modeling

Rahul Kasat, Nien-Hwa Linda Wang, and **Elias Franses**. Purdue University, West Lafayette, IN 47906

The concept of "chirality" is known for almost two centuries. Chiral recognition plays an important role in peptide biosynthesis, amino acid synthesis, drug design, separations, etc. Separations of enantiomers of chiral molecules are important in pharmaceuticals, agrochemicals, and fine chemicals industries. One of the widely used models for explaining the chiral discrimination, the "three point attachment model" was proposed in 1930s for proteins and enzymes. Polysaccharide-based sorbents or biopolymers are used widely in chiral separations because of their versatility in separating wide classes of chiral molecules. The chiral recognition mechanisms of these biopolymers are difficult to determine because of their complex structures. The lack of fundamental understanding of the chiral recognition mechanisms for these systems results in empirical, time consuming, and expensive selection of sorbents in chromatographic enantioseparations.

In this research, the chiral recognition mechanisms for certain important but structurally complex polysaccharides-based sorbents are elucidated [1-4]. The sorbent-solute-solvent interactions are investigated by systematically studying first the sorbent in the dry state [2],and then sorbent-solvent [3], sorbent-non chiral solute [4] and sorbent-chiral solute interactions [1]. A combination of molecular modeling tools, molecular dynamics (MD), and density functional theory (DFT), and experimental techniques, high performance liquid chromatography (HPLC), attenuated total reflection-infrared spectroscopy (ATR-IR), 13C cross polarization magic angle spinning (CP/MAS), and MAS solid state NMR, and X-ray diffraction (XRD) are used.

HPLC is used to measure the retention factors and enantioselectivities [1]. ATR-IR spectra show that the H-bonding states of the key binding sites of the polymer, C=O, and NH, can be different in different sorbents [2]. The DFT-predicted IR results for single polymer side chains help in understanding the shifts in the IR wavenumbers and the subtle changes observed in the intensities in the IR spectra of the dry sorbents [2]. The sorbent H-bonding sites are affected significantly upon absorption of solvents or solutes, providing insights into sorbent-solute and sorbent-solvent interactions [1-4]. The DFT results are consistent with the observed IR shifts and provide measures of the strengths of different interactions (hydrogen bonding and phenyl-phenyl) in different sorbent-solute and sorbent-solvent configurations [1, 3, 4]. The DFT simulations also predict side-chain conformations for different sorbents [2]. From the experimental and the DFT-predicted NMR chemical shifts of the C-1 carbon in the polymer backbone, it is inferred that the backbone glycosidic bond conformations and helicities are similar for amylose-based sorbents and are different from those of cellulose-based sorbents [2]. Solid state NMR studies show that the polymer side chain groups are more mobile than the polymer backbone groups [2, 3]. The XRD spectra of the dry sorbents and of the sorbents upon absorption of solvents and solutes provide structural information on the polymer helical pitch and packing arrangements of these polymers [1-3]. The polymer backbone conformations, as inferred from NMR, the polymer helical pitch, as observed in

XRD, and the polymer side chain conformations from DFT are used to help build the polymer molecular models used for the sorbent-solute MD studies [1-4].

A novel but simple functional group approach is proposed. It allows probing the complex sorbent-chiral solute interactions. The retention times of complex chiral solutes (molecules with 3 or 4 functional groups) are compared with those of simpler non-chiral or chiral solutes (molecules with 2 or 1 functional groups). The results have allowed us to quantify the number and the nature of the sorbent-chiral solute interactions for the first time in these systems. The systematic variations in the structures of the complex chiral solutes have helped us identify for these systems the minimum number of interactions required for the chiral recognition and resulting in a significant enantioselectivity. The elution orders predicted for the enantiomers of the chiral solutes studied using sorbent-solute MD simulations are consistent with the HPLC results. The results support the hypothesis that for a significant enantioselectivity a combination of at least three different interactions are required among three types: H-bonding, phenyl-phenyl, or steric hindrance. Variations in the number and the type of the interactions and in the spatial arrangements of the functional groups in chiral solutes result in significantly different retention factors and enantioselectivities.

In summary, molecular modeling, guided by detailed experimental results, provides valuable insights into molecular details not easily accessible by experiments, and helps elucidate chiral recognition mechanisms.

1. R. B. Kasat, N.-H. Linda Wang, and E. I. Franses, Journal of Chromatography A, 1190, 2008, 110-119.

2. R. B. Kasat, N.-H. Linda Wang, and E. I. Franses, Biomacromolecules, 8, 2007, 1676-1685.

3. R. B. Kasat, Y. Zvinevich, H. W. Hillhouse, K. T. Thomson, N.-H. Linda Wang, and E. I. Franses, Journal of Physical Chemistry B, 110, 2006, 14114-14122.

4. R. B. Kasat, C. Y. Chin, K. T. Thomson, E. I. Franses, and N.-H. Linda Wang, Adsorption, 12, 2006, 405-416.

451f Macromolecular Surfactants: Engineering Value-Added and Commodity Products

Frank Bates, University of Minnesota, Minneapolis, MN 55455

Block copolymers belong to a broad class of amphiphilic compounds that includes lipids, soaps, and nonionic surfactants. A macromolecular architecture affords certain unique advantages over conventional low molecular weight amphiphiles in constructing nanoscale objects with prescribed morphologies and physical properties. I will describe the formation of structures classically found in mixtures of oil, water and surfactant. The static and dynamic properties of spherical and worm-like micelles, vesicles, and bicontinuous microemulsions have been investigated in a variety of systems using small-angle x-ray and neutron scattering (SAXS and SANS), electron and optical microscopy, and in-situ rheological techniques. This presentation will highlight the similarities and differences between macromolecular and traditional amphiphiles, including several illustrative examples that address practical needs in fields ranging from biomedical engineering to structural plastics.

506a On the Role of Inherent Structure Dynamics In Glass-Forming Materials

Dimitrios Tsalikis, Nikolaos Lempesis, Georgios C. Boulougouris, and **Doros N. Theodorou**. School of Chemical Engineering, National Technical University of Athens, 9 Heroon Polytechniou Street, Zografou Campus, Athens, 157 80, Greece

In this work [1-2] we investigate the role of *inherent structures*, i.e., minima on the potential energy hypersurface with respect to all microscopic degrees of freedom, in the vitrification process of glass-forming materials.

For temperatures below the glass temperature T_g we have envisioned a model that describes the dynamics of the atomistic system as a series of transitions, following first-order kinetics and Poisson statistics, between basins surrounding the inherent structures. Our validation of this model is based on its ability to reproduce the mean square atomic displacement resulting from a full-blown atomistic molecular dynamics (MD) trajectory. The *projection* of MD trajectories onto the Poisson process model has been made possible thanks to the use of hazard plot analysis, which allowed the evaluation of the rate constants for individual transitions between the basins. Our results for a model two-component Lennard-Jones mixture [3-4] clearly show that our Poisson process approach is able to reproduce the inherent structure dynamics obtained directly from MD over a wide range of sub-T_g temperatures, from 9K to 38 K. At higher temperatures, the long-time diffusive motion of atoms in the system can be captured simply by tracking the mean square displacement between inherent structures.

We utilize and extend this methodology in order fully to reconstruct the system dynamics in the form of the mean square atomic displacement as a function of time at finite temperatures based on the succession of transitions in a network of basins or "states," the contribution of vibrational motion around each inherent structure being evaluated through short MD runs artificially trapped within each one of the states. To quantify the vibrational contribution we propose a procedure that traps the system within each of the basins by applying reflective conditions, that is, reversing the atomic momenta in Newton's equations of motion every time the minimization procedure shows that a switch to the basin of a neighboring inherent structure has occurred. The inversion of the atomic momenta is followed by a randomization of the momentum of the thermostat that smoothly leads to a new trajectory. This procedure has been inspired by the novel methods developed by Voter [5] for analyzing the dynamics of infrequent events, where reflective conditions were introduced in the context of biasing the local search for transitions, by increasing the temperature.

We provide [2] the mathematical formulation for *lifting* the coarse-grained Poisson process model of transitions between states back to the atomistic level and thereby reproducing the full dynamics of the atomistic system within the Poisson approximation. Our results for the mean square atomic displacement as a function of time show excellent agreement with the full MD for temperatures around and below the glass temperature of our model Lennard-Jones two-component mixtures. Clearly, our approach is able to reproduce the full dynamics of the atomistic system at low temperatures, where the Poisson approximation is valid.

The proposed methodology creates a network analogous to a network of first-order, reversible chemical reactions, with basins or states around the inherent structures playing the role of chemical species. When the system remains trapped in a small number of inherent structures, as happens at temperatures lower than the glass temperature, it is possible to derive an analytical solution for the evolution of the probability distribution of the system among states, given the transition rate constants. The ability to identify new states to be accessed and calculate the transition rate constants on the fly can be invoked in order to build an ever-expanding network of states and use it to evaluate analytically averages over all possible dynamic outcomes of the Poisson process up to a given observation time. This has been achieved by Boulougouris and Theodorou [6] via the development of *Dynamic Integration over a Markovian Web* (DIMW). This method relies upon a distinction between "explored" states and "boundary" states and invokes absorbing boundary conditions to evaluate analytically the mean first passage time to reach any of the boundary states. The augmentation of the network (by appending a boundary state to the set of explored states) is accomplished by a stochastic selection scheme based on the flux to each of the boundary states.

Our ultimate aim is to develop a methodology that enables mapping the dynamics around and below T_g to a coarse-grained first-order kinetic model based on the Poisson process approximation. To do so, we are combining our MD-based approach with the DIMW method [6]. This will allow us not only to observe the time evolution of the system, but also to sample over a time range far beyond the accessible times of classical MD. Therefore, it will allow us to create computer glasses at cooling rates closer to those used experimentally.

References

[1] Tsalikis D. G., Lempesis N., Boulougouris G. C. and Theodorou D. N., submitted.

[2] Tsalikis D. G., Lempesis N., Boulougouris G. C. and Theodorou D. N., submitted.

[3] Kob, W.; Andersen, H. C. *Phys. Rev. Lett.* **1994**, *73*, 1376.

[4] Kob, W. *Phys. Rev. E.* **1995**, *51*, 4626.

[5] Voter, A. F. *Phys. Rev. Lett.* **1997**, *78*, 3908; Voter, A. F.; Doll, J. D. *J. Chem. Phys.* **1985**, *82*, 80.

[6] Boulougouris, G. C.; Theodorou, D. N. *J. Chem. Phys.* **2007**, *127*, 084903.

Web Page: comse.chemeng.ntua.gr

506b Novel Monte Carlo Algorithm to Sample Intramolecular Degrees of Freedom of Linear, Branched and Cyclic Molecules with Fixed Bond Length Constraints

Jindal K. Shah and Edward J. Maginn. Department of Chemical and Biomolecular Engineering, University of Notre Dame, 148 Fitzpatrick, Notre Dame, IN 46556

Efficient and correct sampling of bond angle distributions for molecular models with fixed bond length constraints present considerable challenges in Monte Carlo simulations. Most of the algorithms developed to date address the issue by focusing on internal coordinate generation and thus are difficult to implement in an existing Monte Carlo code. In this presentation, we demonstrate an exceedingly simple and efficient MC algorithm that correctly generates bond angle distributions at a branch point. This is achieved by perturbing Cartesian coordinates of a randomly chosen atom of the branch point. The algorithm is designed to preserve the fixed bond length constraint at each step. Results obtained from the algorithm for bond angle distributions for branch points containing three and four branches have been validated against those produced by a brute force Boltzmann rejection scheme at 298 and 1000 K. Extension of the method in combination with ring closure will also be discussed for efficient sampling of cyclic molecules with constrained bond lengths such as an all-atom model of benzene.

506c Optimizing Sampling and Staging for Simulations of Rare Events Via Forward Flux Sampling Schemes

Ernesto E. Borrero-Quintana, School Chemical and Biomolecular Engineering, Cornell University, 120 Olin Hall, Ithaca, NY 14853 and Fernando Escobedo, School of Chemical and Biomolecular Engineering, Cornell University, 120 Olin Hall, Ithaca, NY 14853.

Forward flux sampling (FFS) schemes have become a valuable tool for studying the dynamics of complex systems; however, its applicability is highly limited by sampling efficiency, which is dictated primarily by the choice of the order parameter but also by how the phase space is partitioned and sampled. To address these shortcomings, we propose a simple adaptive optimization algorithm which optimizes the phase space staging by concentrating the sampling in the "bottlenecks" of the FFS-type simulation. The algorithm is based on an iterative procedure that systematically identifies the kinetic bottlenecks along the reaction coordinate (i.e., order parameter used to partition the phase space via interfaces) and improves the estimates of the conditional probabilities to reach subsequent interfaces (P_i) to then improve the sampling of the phase space. Hence, the approach can optimize for either the number and position of the interfaces (i.e., optimized λ phase staging), and/or the number M of fired trial runs per interface (i.e., the $\{M_i\}$ set) to minimize the statistical error in the rate constant estimation per simulation period (i.e., increase the computational efficiency). The method is demonstrated for the reassembly and folding in split proteins systems that are representative of numerous protein-protein interactions encountered in many biological applications. The kinetic studies of such systems by FFS-type simulations are usually limited by the choice of a suitable reaction coordinate (i.e., a model on relevant variables) to describe the intrinsic system's dynamics. It is shown that the proposed approach leads to an optimized λ staging and/or $\{M_i\}$ set which optimize the flux of partial trajectories between interfaces based on an input desirable distribution of P_i values. Indeed, the λ phase staging could be optimized first to position interfaces at the transition state and FFS's "bottleneck" regions and the $\{M_i\}$ set then be optimized to improve the sampling and reduce the effort needed to obtain committor probability data for the estimation of refined reaction coordinate models. When taken together, these algorithms provide a gain in efficiency of over two orders of magnitude when compared to traditional FFS simulations.

506d Simulating Nucleation. A New Approach to Overcome the Free Energy Barrier

Isamu Kusaka, Chemical and Biomolecular Engineering, The Ohio State University, Room 125A, Koffolt Labs., 140W. 19th Ave., Columbus, OH 43210

We present a new methodology to study nucleation by means of molecular simulation. A conventional approach in the field has been to use umbrella sampling technique, in which a bias potential given in terms of a set of order parameters forces the system to visit high free energy states, including the one containing a critical nucleus, so that the free energy barrier of nucleation can be evaluated accurately. This approach, however, requires one to identify order parameters that can describe nucleation.

During the course of our study of bubble nucleation, we have tested many order parameters and found that they cannot control the free barrier crossing in a reversible fashion, rendering impossible the accurate evaluation of free energy barrier by this method. In fact, identification of a proper set of order parameters presents a considerable challenge in the conventional use of umbrella sampling technique.

The new approach we developed still uses the umbrella sampling technique, but only to drive the system out of the metastable free energy well. Thus, the bias potential is made to vanish well before the system reaches critical nucleus, and hence has no effect on nucleation behavior except to greatly accelerate the barrier crossing. A method has been devised to evaluate the unbiased first passage time, i.e., the time required for the system to cross over the free energy barrier in the absence of biasing potential, from the biased first passage time evaluated by simulation. We verified that the unbiased first passage time thus obtained is independent of the details of the imposed bias.

Our current implementation of the method is through Monte Carlo method, which allows for extraction of purely thermodynamic information on nucleation behavior decoupled from dynamical aspects of the process. However, we note that the method can also be implemented with molecular dynamics to yield dynamical information.

Results for both bubble nucleation and vapor phase condensation will be reported.

506e Statistically Optimal Free Energy Estimates from Sparsely Chosen States

Michael R. Shirts, Chemical Engineering, University of Virginia, 102 Engineers Way, P.O. Box 400741, Charlottesville, VA 22904-4741 and John D. Chodera, Department of Chemistry, Stanford University, Chemistry Operations MC 5080, Stanford, CA 94605-5080.

We recently presented a minimum variance method to calculate free energies and ensemble averages from multiple equilibrium simulations conducted at different thermodynamic states [1]. This method, based on statistical techniques for correcting biases in sampling, is useful for analyzing both simulations of phase changes and the results of single molecule experiments. This estimator gives the same results as multiple histogram methods [2,3] in the limit of vanishingly small histogram bins, but does not actually require histograms, eliminating bias due to binning. This new estimator is especially useful for highly multidimensional problems where histograms cannot be well populated. Unlike multiple histogram techniques, it also yields an expression for the statistical uncertainty of the estimates it produces.

However, this estimator (like multiple histogram techniques) requires reevaluation of the potential energy of the system at all other thermodynamic states under study. This can become a computationally unreasonable burden when there are many states, such as 3-dimensional potentials of mean force obtained through umbrella sampling or alchemical simulations with many intermediates, or in situations where the cost of obtaining the extra energy evaluations are high, such as when using polarizable force fields. This burden is especially unreasonable since many states with low overlap will contribute little to the corresponding free energy or ensemble averages. Methods such as Bennett's acceptance ratio [4,5] can be used to compute free energies between only neighboring, but this type of pairwise analysis ignores correlation between samples and is inappropriate when states overlap in phase space with many others.

To solve the problem of efficiently computing averages and free energies with large numbers of states, we derive a modification of the multistate minimum variance method that uses only states with sufficient mutual phase space overlap. For many cases of interest, this drastically decreases the number of states that must be considered with negligible loss of precision. We also present a number of test cases of this method, including free energy profiles for single molecule pulling experiments and the computation of small molecule partition coefficients.

[1] M. R. Shirts and J. D. Chodera, http://arxiv.org/abs/0801.1426 (2008)

[2] A. M. Ferrenberg and R. H. Swendsen, Phys. Rev. Lett. 63:1195-1198 (1989)

[3] S. Kumar, D. Bouzida, R. H. Swendsen, P. A. Kollman and J. M. J. Rosenberg, J. Comput. Chem., 13:1011-1021 (1992)

[4] C. H. Bennett, J. Comput. Phys. 22:245-268 (1976)

[5] M. K. Fenwick and F. A. Escobedo, J. Chem. Phys. 120:3066-3074 (2004)

506f Fast off-Lattice Monte Carlo Simulations of Polymeric Systems

Yuhua Yin and **Qiang Wang**. Chemical and Biological Engineering, Colorado State University, Fort Collins, CO 80523

Conventional molecular simulations of multi-chain systems are hindered by harsh excluded-volume interactions and expensive pair-potential calculations. The former greatly reduces the chain relaxation towards equilibrium configurations and the efficiency of sampling the configurational space, while the latter becomes computationally very expensive for concentrated polymeric systems.

The idea of fast off-lattice Monte Carlo (FOMC) simulations is to perform particle-based Monte Carlo simulations in continuum with a Hamiltonian commonly used in polymer field theories such as the well developed self-consistent field (SCF) theory, where individual polymer segments are modeled as volumeless points and the excluded-volume interactions are modeled by the Helfand compressibility. This avoids the harsh repulsion and, in some cases, the pair-potential calculations, thus allowing much faster and better sampling of the configurational space. Furthermore, using the same Hamiltonian in both SCF calculations and FOMC simulations enables quantitative comparisons between the two without any parameter-fitting.

To demonstrate the great advantages of FOMC simulations, we have studied two polymeric systems: confined homopolymer solutions and microphase separation of diblock copolymers. In both cases, the FOMC simulations are compared with SCF calculations to reveal the effects of fluctuations ignored in the latter.

506g Molecular Dynamics In the Isothermal-Isobaric Ensemble: Discontinuous Potentials

Mark J. Uline and **David S. Corti**. School of Chemical Engineering, Purdue University, 480 Stadium Mall Drive, West Lafayette, IN 47907

Based on the approach of Gruhn and Monson [*Phys. Rev. E*, **63**, 061106 (2001)], we present a new method for deriving the collisions dynamics for particles that interact via discontinuous potentials. By invoking the conservation of the extended Hamiltonian, we generate molecular dynamics (MD) algorithms for simulating the hard-sphere and square-well fluids within the isothermal-isobaric (*NpT*) ensemble. Consistent with the recent rigorous reformulation of the *NpT* ensemble partition function, the equations of motion impose a constant external pressure via the introduction of a shell particle of known mass [Uline and Corti, *J. Chem. Phys.*,**123**, 164101 and 164102 (2005)], which serves to define uniquely the volume of the system. The particles are also connected to a temperature reservoir through the use of a chain of Nosé-Hoover thermostats, the properties of which are not affected by a hard-sphere or square-well collision. By using the Liouville operator formalism and the Trotter expansion theorem to integrate the equations of motion, the update of the thermostat variables can be decoupled from the update of the positions of the particles and the momenta changes upon a collision. Hence, once the appropriate collision dynamics for the isobaric-isenthalpic (*NpH*) equations of motion are known, the adaptation of the algorithm to the *NpT* ensemble is straightforward. Results of MD simulations for the pure component square-well fluid are presented and serve to validate our algorithm. Finally, since the mass of the shell particle is known, the system itself, and not a piston of arbitrary mass, controls the time scales for internal pressure and volume fluctuations. We therefore consider the influence of the shell particle algorithm on the dynamics of the square-well fluid.

On the Computation of Spectra of Spatially Varying Convective-Diffusion Operators

By

Ansgar Bohmann[1,2], Nandkishor Nere, Doraiswami Ramkrishna[1]* & Achim Kienle[2]

[1]School of Chemical Engineering, Purdue University, West Lafayette, IN 47907
[2]Max Planck Institute for Dynamics of Complex Technical Systems; Otto-von-Guericke-Universität, Magdeburg, Germany

Numerous applications in chemical engineering involve linear problems with spatially varying transport coefficients whose solutions can be obtained by spectral methods using the eigenvalues and eigenvectors of the associated Sturm-Liouville operators. Accurate computation of the spectra of Sturm-Liouville operators with non-constant coefficients require accurate numerical methods. In this connection, we demonstrate in this work, the effectiveness of the so-called Prufer transformation with respect to both computational speed and accuracy in comparison with Matlab methodology.

Motivation

Spatially varying problems arise naturally in turbulent transport. Consider, for example, the break-up of liquid droplets in turbulent flow of a continuous phase in which the droplets are immiscible. Assuming axisymmetry, variations occur in quantities of interest along the radial and axial directions. The population balance equation in the number density of drops in the i th size range, denoted $N_i(r,z)$ satisfies the following equation

$$Pe u_z(r)\frac{\partial N_i(r,z)}{\partial z}-\frac{1}{r}\frac{\partial}{\partial r}\left(r\mathcal{D}(r)\frac{\partial N_i(r,z)}{\partial r}\right)$$
$$=\beta\Gamma_R(r)\left[-\Gamma_X^{(i)}N_i(r,z)+\sum_{k=i}^{n}\gamma_{i,k}\Gamma_X^{(k)}N_k(r,z)\right],\quad i=1,2,\ldots,n \tag{1}$$

The details of the formulation are available in Ramkrishna and Nere (2007). The foregoing equation may be rewritten in more compact vector form as

$$Pe u_z(r)\frac{\partial \mathbf{N}(r,z)}{\partial z}-\frac{1}{r}\frac{\partial}{\partial r}\left(r\mathcal{D}(r)\frac{\partial \mathbf{N}(r,z)}{\partial r}\right)=\beta\Gamma_R(r)\mathbf{A}\mathbf{N}(r,z) \tag{2}$$

The boundary conditions may be written for $\mathbf{N}$ as

$$\frac{\partial \mathbf{N}}{\partial r}=\mathbf{0},$$

The solution of this problem can be readily obtained using the spectral decomposition of the Sturm-Liouville operator

The Sturm Liouville Operator

The Sturm-Liouville operator is given by

$$L\equiv-\frac{1}{w(x)}\frac{d}{dx}\left[p(x)\frac{d}{dx}\right]+q(x),\qquad a<x<b \tag{3}$$

with homogeneous boundary conditions defining a domain $D(L)$ given by

$$u(a)\cos\alpha - p(a)u'(a)\sin\alpha = 0, \qquad u(b)\cos\beta - p(b)u'(b)\sin\beta = 0 \qquad (4)$$

for which the operator $\mathbf{L} = \{L, D(L)\}$ is self-adjoint. In particular note that p,q,w depend on x and are positive. The eigenvalues and eigenvectors, obtained by solving $\mathbf{Lu} = \lambda\mathbf{u}$, can be used to solve boundary value problems involving the operator $\mathbf{L}$. The solution of the differential equation $Lu(x) = \lambda u(x)$ provides $\mathbf{u}(\lambda) \equiv \{u(x,\lambda)\}$. The eigenvalues are obtained by demanding that $\mathbf{u}(\lambda) \in D(L)$. The problem of accurately determining the eigenvalues and eigenvectors can be difficult for the cases where p,q,w depend on x.

The Prufer Transformation

The Prufer transformation is as follows. Defining $u' = \frac{1}{p}v$, $v' = (q - \lambda w)u$, where primes denote differentiation with respect to x, we define polar coordinates, $v = r\cos\theta$, $u = r\sin\theta$. For nontrivial solutions, the differential equation $Lu = \lambda u$ leads to the pair of first order differential equations

$$\theta' = \frac{1}{p}\cos^2\theta + \lambda(w - q)\sin^2\theta, \quad \frac{r'}{r} = \left(\frac{1}{p} - \lambda w + q\right)\sin\theta\cos\theta \qquad (5)$$

with boundary conditions $\theta(a) = \alpha, \quad \theta(b) = \beta$ obtained by transforming (4). Since both boundary conditions involve only the function θ and the (nonlinear) differential equation for θ is independent of r, the differential equation in r may be integrated to obtain

$$\ln|r(x)| = \int_{x_o}^{x}\left(\frac{1}{p(x')} - \lambda w(x') + q(x')\right)\sin\theta(x')\cos\theta(x')\,dx' \qquad (6)$$

where it is presumed that r does not vanish in the interval $a \le x \le b$. Since both boundary conditions are given in terms of only the function θ the calculation of eigenvalues involves only the integration of (5). The details concerning the calculation of the eigenvalues, available in a dissertation by Bohmann (2008), will be briefly included in the presentation. Broadly, however, the method involves the use of a shooting method to match the solution at some intermediate point c in the interval (a,b). Several academic examples were solved using the method, one of which is included in this abstract.

$$u'' + e^x u = \lambda u, \quad 0 < x < \pi; \qquad u(0) = u(\pi) = 0$$

The results of calculations are shown alongside for the eigenvalues and eigenvectors. The paper will discuss the methodology for solving the problem of drop break-up in fully developed flow of a liquid-liquid dispersion.

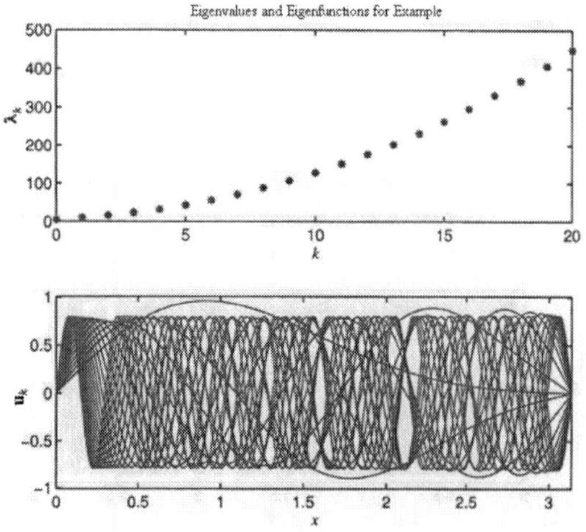

References

Bohmann, A., "Eigenvalue methods and the Prufer Transformation in Chemical Engineering," B.S. Thesis submitted to the University of Magdeburg and Purdue University, July 2008.

564b Adsorption and Interfacial Consolidation of Amphoteric Species at Oil-Water Interfaces

Vincent J. Verruto[1], Rosemary K. Le[1], and **Peter Kilpatrick**[2]. (1) Chemical and Biomolecular Engineering, North Carolina State University, 911 Partners Way, Engineering Building 1, Rm 2-102, Raleigh, NC 27695-7905, (2) University of Notre Dame, Notre Dame, IN 46556

The formation of stable emulsions through the adsorption of polyvalent, complex naturally occurring species is important in a variety of important industrial applications and challenges, including proteins and other macromolecules in food emulsions, and asphaltenes in petroleum and bitumen-based emulsions. The stability of these emulsions is attributed to interfacial films with viscoelastic properties that are known to vary with concentration, solvent quality, and macromolecule chemistry. Here, we explore the impact of aqueous phase chemistry, namely pH and salinity, on the transient interfacial rheological properties of asphaltenic films. With two chemically unique asphaltenes, interfacial shear rheology revealed salt-induced retardation of the interfacial consolidation process that ultimately engenders elasticity to the film. With HO asphaltenes at pH 7, a linear dependence of this retardation on the Debye parameter (κ) indicated that shielding of electrostatic attraction was responsible. Further investigation with dynamic oscillating drop tensiometry at varying pH, to probe dilatational rheological properties as well as dynamic interfacial tension, illustrated that electrostatic interactions, both attractive and repulsive, clearly influence the evolution of the interfacial structure. More specifically, the transient tension and dilatational moduli indicated that several interfacial processes were impacted by the addition of salt, including (i) interfacial activity and the extent of adsorption, (ii) interfacial rearrangement and consolidation, and (iii) interfacial transport and displacement. We believe that these observations are likely to be universal with other macromolecular systems of amphoteric character. We further elaborate on this by considering the globular proteins lysozyme and β-casein, and their adsorption and consolidation at oil-water interfaces.

564c Mixtures and Nanostructures in Enhanced Oil Recovery

George J. Hirasaki, Rice University, Houston, TX 77005-1892 and Clarence Miller, Chemical and Biomolecular Engineering, Rice University, 6100 Main Street, Houston, TX 77005-1892.

Fluids utilized in Enhanced Oil Recovery (EOR) often have complex nanostructure in the process of achieving ultra-low interfacial tension (IFT) and/or mobility control. Examples are illustrated from alkaline surfactant polymer (ASP) flooding and carbon dioxide foam.

Synergism can occur with blending of two dissimilar surfactants. N67-7PO sulfate is a C16-C17 slightly branched propyoxylated sulfate surfactant that is tolerant of divalent ions. IOS15-18 is a C15-C18 internal olefin sulfonate that is randomly branched but intolerant of divalent ions. Mixtures of these two surfactants have a single-phase region with either NaCl or CaCl2 that is larger than with either surfactant alone.

Naphthenic soaps are generated in situ as a co-surfactant in ASP flooding. These soaps are generally more lipophilic that the usual sulfate and sulfonate surfactants used in EOR. This results in the optimal salinity for minimum IFT being a function of the soap/surfactant ratio or fraction. In addition, it was discovered that in under-optimum conditions (Winsor I), the lower-phase microemulsion is not a homogeneous phase. Rather it has a composition gradient with a more oil-rich microemulsion (colloidal dispersion) in contact with the excess-oil phase at gravity equilibrium. This oil-rich microemulsion is a middle layer with respect to the lower-phase microemulsion and excess oil. If this oil-rich microemulsion is present to coat the drop of excess-oil in a spinning-drop measurement, the IFT is ultralow (below 10-2 mN/m) over a much wider salinity range than in its absence.

Polymer is included in the surfactant slug of the ASP process for mobility control. If the salinity of the injected surfactant slug is near the optimal salinity of the pure surfactant, the solution may separate into surfactant-rich and polymer rich phases. This can result in loss of mobility control. However, the surfactant slug can be injected well below the optimal salinity of the pure surfactant and the ultra-low IFT region generated in situ as the surfactant mixes with the naphthenic soaps.

Carbon dioxide is widely used for EOR but often has poor sweep efficiency because of its low viscosity of about 0.03 cp. However, CO2 can have much higher apparent viscosity as it flows through porous media if it becomes dispersed in water as foam (or emulsion if below Tc). There are surfactants that can be injected solubilized in CO2 but yet generate in situ a CO2 in water foam (emulsion) as it mixes with water in the formation.

564d Surface Nanostructure, Diffusion and Catalysis: The Role of Confinement and Surface Chemistry

Keith Gubbins, North Carolina State University, Raleigh, NC 27695

Chemical reactions are often carried out in nano-structured materials, which can enhance reactions due to interactions with the reacting mixture, high surface area and confinement effects. A systematic study of reactions in confinement is presented, using ab initio and semi-classical methods [1]. We first consider the influence of steric hindrance on the equilibrium and kinetics for the rotational isomerizations of small hydrocarbons, and show that confinement can lead to reaction rates that vary doubly exponentially with pore width [2]. We further consider several reactions to produce hydrogen, including the decomposition of methane and water on carbon surfaces, with and without the presence of defects and added transition metal atoms. These results provide examples of the influence on reaction mechanism, yield and rate of electrostatic interactions with the supporting material, surface defects and surface curvature.

[1] Erik E. Santiso, Milen K. Kostov, Aaron M. George, Marco Buongiorno Nardelli, Keith E. Gubbins, "Confinement Effects on Chemical Reactions – Towards an Integrated Rational Catalyst Design", Applied Surface Science, 253, 5570 (2007).

[2] E.E. Santiso, M. Buongiorno Nardelli and K.E. Gubbins, "A Remarkable Shape-Catalytic Effect of Confinement on the Rotational Isomerization of Small Hydrocarbons", Journal of Chemical Physics, 128, 034704 (2008).

564e Simulating Very Crowded Pores with Precise Detail: Ted Takes on Zeolites

Alon McCormick, University of Minnesota, Minneapolis, MN 55455

I will review some of H. Ted Davis' contributions to molecular theory that have found direct use in understanding trends of molecular selectivity in zeolite adsorption and diffusion. I will end with a short overview of Ted's true legacy in this field - the contributions made by some of his former students who have become very active in zeolites, molecular sieves, and molecular shape selectivity.

564f Dispersive Flow In Beadpacks

H. Ted Davis[1], Daniel Kroll[2], Robert Maier[3], Alon McCormick[1], and E. Vandre[1]. (1) University of Minnesota, Minneapolis, MN 55455, (2) University of North Dakota, Fargo, ND 58108, (3) U.S. Army Engineering Research and Development Center, Vicksburg, MS 39180

In computer simulations we carried out of dispersion in fluid flow in cylindrical beadpacks, it appeared that the longitudinal dispersion coefficient, DL, was dependent on the ratio of the cylinder radius to bead diameter (R/d) much more than was indicated in the literature, where it was postulated that for R/d greater than 5 the dispersivity would be independent of R/d. Since the computer simulations were limited to relatively small systems, one couldn't be certain that our results were not the result of a small sample approximation. Thus, we investigated dispersion using the approach Aris used in his analysis of Taylor dispersion in tube flow. Predictions of this analysis indicated a strong dependence of dispersivity on R/d at ratios well above R/d=5. This last summer, we carried out experiments on beadpacks with R/d = 9.5, 12.5 and 25.4. We found a strong dependency of DL on R/d, in close agreement with the analysis.

596a First-Principles Simulations of Noble Gases Dissolved in Liquid Silica

Liqun Zhang[1], Dan Lacks, and James Van Orman[2]. (1) Chemical engineering and Geology Science, Case Western Reserve University, K. W. Smith building, 10900 Euclide Avenue, Cleveland, OH 44106, (2) Geology Science, Case Western Reserve University, K. W. Smith building, 10900 Euclide Avenue, Cleveland, OH 44106

Noble gases are widely used in geochemistry to gain insight into the structure of the deep Earth, its formation and differentiation, and the efficiency of large-scale processes such as mantle convection, These studies depend in part on understanding the solubility and local structure of noble gases in Earth materials, particularly silicate liquids, under high pressures and temperatures. Silicate liquids undergo profound changes in structure with increasing pressure, and it is not well known how dissolved noble gases respond to these modifications.

First principles molecular simulations were performed, using VASP, on five different noble gases in liquid silica at 4000 K, under pressures from ambient pressure to around lower mantle pressure. The sizes of noble gas atoms were estimated based on radial distribution function results, and were found to vary linearly with the volume ratio of the simulated liquid to its ambient melt. The VASP results agree well with available experimental results for Kr and should be more accurate than former available estimations.

The number of nearest-neighbor oxygen atoms surrounding a noble gas atom was also investigated. At high pressures, where silica adopts a more close-packed structure, the coordination number was found to be in excellent agreement with theoretical results on the number of hard spheres of different size that can be packed around a central sphere.

596b First Principles Simulations of Aqueous Systems

J. Ilja Siepmann[1], Kelly E. Anderson[2], Matthew J. McGrath[2], I.-F. Will Kuo[3], and Christopher J. Mundy[4]. (1) University of Minnesota, Department of Chemistry, Minneapolis, MN 55455-0431, (2) Department of Chemistry, University of Minnesota, 207 Pleasant St. SE, Minneapolis, MN 55455, (3) CMS, Lawrence Livermore National Laboratory, East Avenue, Livermore, CA 94550, (4) Fundamental and Computational Sciences Directorate, Pacific Northwest National Laboratory, Richland, WA 99352

First principles molecular dynamics and Monte Carlo simulations with the electronic structure program CP2K are carried out to investigate (i) the double ion formation in small clusters consisting of water, sulfuric acid, and a nitrogen base, (ii) the speciation in hydrous silica mixtures at Earth mantle conditions, and (iii) the solubility of hydrogen chloride in water.

596c An Approach for Developing Intermolecular Models for Use within Saft-Vr from Quantum Mechanical Calculations and Experimental Data

Beatriz Giner[1], Timothy Sheldon[1], Michaela Pollock[1], **Claire S. Adjiman**[1], Amparo Galindo[1], George Jackson[1], Denis Jacquemin[2], Valerie Wathelet[2], and Eric A. Perpete[2]. (1) Department of Chemical Engineering, Imperial College London, Centre for Process Systems Engineering, London, SW7 2AZ, United Kingdom, (2) Chemistry, Facultes Universitaires Notre Dame de la Paix, Rue de Bruxelles, 61, Namur, 5000, Belgium

Molecular equations of state such as SAFT (statistical associating fluid theory) and its variants [1] are particularly well suited to describing the fluid phase behavior of complex systems such as polymers and hydrogen-bonding compounds. However, such approaches require a minimum of three parameters for each pure compound to be modeled, and additional parameters for mixtures. Their use is thus dependent upon the availability of reliable parameter values for the compounds of interest.

SAFT parameters for pure components are usually obtained using fluid phase equilibrium data. In order to obtain statistically significant values of the parameters, a large number of experimental data points are required. Furthermore, it is desirable to include different types of data, such as saturated vapor pressures and saturated liquid densities to allow the resolution of different parameters: the density is known to be most sensitive to size parameters, while the vapor pressure depends most strongly on energy parameters. Even when a large and varied data set is available, some of the parameter values remain difficult to determine with precision. In SAFT-like methods, this is the case of the chain-length (aspect ratio) parameter, m, which is often fixed a priori based on physical arguments, rather than fitted to experimental data [e.g., 2]. When using SAFT with variable range (SAFT-VR), the identification of the range parameter can be made more reliable by including additional data such as the speed of sound [3]. Finally, when association is present, it can be difficult to partition the attractive interactions between the dispersive and associating terms and data such as the fraction of bonded molecules can be useful [4].

In practice, there are few compounds for which extensive experimental data are available. It is therefore desirable to look for alternative approaches which provide reliable parameter values while reducing the dependence on experimental data. Several efforts have been made in this direction, since the pioneering work of Wolbach and Sandler [5]. Approaches based on quantum mechanics and/or molecular simulations have been developed and applied to small sets of compounds [e.g., 6].

In this work, we focus on deriving the size parameters of the SAFT-VR equation of state *ab initio*, making sure that the approach is applicable to different types of compounds. We present two approaches. In the first approach, only the chain-length parameter, m, is obtained from quantum mechanics calculations at the Hartree-Fock level of theory, and the remaining parameters are fitted to vapor pressure and saturated density data. In the second approach, both m and s are obtained from quantum mechanics, and the remaining parameters are estimated from data. We apply these two procedures to a wide range of compounds, including long-chain and associating compounds, for which the SAFT-VR equation is well-suited. The test set of compounds comprises *n*-alkanes, cyclic and aromatic compounds, small gases, refrigerants and refrigerant intermediates. We show that high quality results can be obtained by following this approach in terms of the average error on vapor pressure and saturated density. The models derived are consistent with physical trends, for instance in terms of carbon number. This provides confidence that the models can be successfully used in modeling mixture behavior. This is demonstrated on a few sample mixtures.

The applicability of the approach when no density data are available is also investigated. In this case, the values of *m* and *s* derived from quantum mechanics prove particularly useful: it is shown that the SAFT-VR models thus developed can be used to predict saturated densities in the absence of such data. If *s* is instead obtained by fitting to vapor pressure data only, physically unrealistic values of *s* are derived and the density predictions are very poor.

1. W.G. Chapman, K.E. Gubbins, G. Jackson and M. Radosz, Fluid Phase Eq. 52 (1989) 31; W.G. Chapman, K.E. Gubbins, G. Jackson and M. Radosz, Ind. Eng. Chem. Res. 29 (1990) 1709; E.A. Müller and K.E. Gubbins, Ind. Eng. Chem. Res. 40 (2001) 2193.

2. C. McCabe and G. Jackson, Phys. Chem. Chem. Phys. 1 (1999) 2057; P. Paricaud, A. Galindo and G. Jackson, Ind. Eng. Chem. Res. 43 (2004) 6871.

3. T. Lafitte, D. Bessieres, M.M. Piñeiro, J.-L. Daridon, J. Chem. Phys. 124 (2006) 024509.

4. G. Clark, A. J. Haslam, A. Galindo, and G. Jackson, Mol. Phys., 104 (2006) 3561.

5. J.P. Wolbach and S.I. Sandler, Ind. Eng. Chem. Res. 36 (1997) 4041.

6. J.P. Wolbach and S.I. Sandler, Int. J. Thermophysics 18 (1997) 1001; M. Yarrison and W.G. Chapman, Fluid Phase Eq. 226 (2004) 195; M. Fermeglia and S. Pricl, Fluid Phase Eq. 166 (1999) 21; M. Fermeglia, A. Bertucco, D. Patrizio, Chem. Eng. Sci. 52 (1997) 1517.

596d DFT Calculation of ^{29}Si-^{1}H Indirect Spin-Spin Coupling Constants In Organoalkoxysilanes

Jyothirmai Ambati and Stephen E. Rankin. Chemical and Materials Engineering Department, University of Kentucky, 177 Anderson Hall, Lexington, KY 40506

To study the reactions of organoalkoxysilanes and their hydrolyzed products, polarization transfer NMR experiments such as Distortionless Enhancement by Polarization Transfer (DEPT) are frequently employed for signal enhancement and spectral editing. However, even seemingly simple precursors such as bis(triethoxysilyl)ethane show complex non-first-order coupling patterns that cannot be interpreted without the aid of first-principles calculations. Here, we discuss the development of methodology for predicting indirect spin-spin coupling constants in organoalkoxysilanes. The performance of four basis sets: 6-31G, 6-311+G(2d,p), cc-PVTZ and IGLO-III is compared regarding the estimation of ^{29}Si-^{1}H spin-spin coupling constants in methyltriethoxysilane. DFT calculations with the 6-31G basis set and B3LYP functional are found to offer good agreement with the experimental coupling constants. This model chemistry is used to calculate the ^{29}Si-^{1}H coupling constants of MexSi(OR)y(OH)4-x-y (OR=OMe, OEt) series of silanes. It is observed that substituting alkoxy groups with hydroxyl groups has no effect on the two bond ^{29}Si coupling to ^{1}H in non-hydrolyzable organic group. Hence, the fully hydrolyzed version of the corresponding monomers can be used to estimate these values. This reduces the computational cost without reducing the accuracy of the results. Based on this approach, ^{29}Si-^{1}H scalar coupling constants in bridged alkoxysilane of interest for kinetic studes are discussed: bis (trimethoxysilyl)ethane, bis(triethoxysilyl)methane, bis(triethoxysilyl)ethane and bis(triethoxysilyl) ethene. The fully coupled ^{29}Si NMR spectral patterns of these molecules, specifically those with ethylene bridging groups, are too complex to directly measure the coupling constants based on peak splitting. We estimated these values with the aid of NMR spin system simulations and the scalar coupling constants from DFT calculations. Coupling constants of the trans conformation provide consistent predictions with the selectively decoupled ^{29}Si spectra, and are unaffected by variations of temperature and solvent.

596e First Principles Monte Carlo Simulations of Elemental Fluid Phase Equilibria

Lev D. Gelb, Department of Chemistry, Washington University in St. Louis, One Brookings Drive, St. Louis, MO 63130

We use modern Monte Carlo sampling methods in concert with standard electronic structure codes (CPMD and NWChem) to perform liquid-vapor and liquid-liquid coexistence calculations from first principles. In order to improve computational efficiency an approximate "pre-sampling" potential is used to generate large moves with a high probability of acceptance, though the actual distribution of sampled states is determined only by the first-principles energy calculations [1].

The liquid elements, especially in the P-block, display a variety of liquid-liquid transitions (most famously in phosphorous [2] and sulfur [3], but also in selenium and others [3]) in addition to normal vaporization equilibria, as well as unusual liquid structure in many cases. All of this behavor is due to the strongly-directional and many-body covalent nature of interactions in these systems. While there has been much recent progress in developing empirical potentials for such chemically reactive systems [4], the use of "potential-less" first-principles methods is particularly appealing in studies of these systems; the relatively high temperatures considered also make it more likely that the known deficiencies in modern density functional methods (poor or inconsistent treatment of dispersion forces, etc.) are less likely to substantially affect the results.

Issues of efficiency, correctness and load-balancing arising from the use of iterative density functional theory potentials are addressed both directly and through analysis of model systems. System size effects are also tested using model systems. A hybrid approach in which two "levels" of pre-sampling is constructed, in which the first uses an empirical potential, and the second uses density functional theory but with a very low (and fixed) number of plane waves. Construction of appropriate pseudopotentials and choice of exchange-correlation functional are evaluated in detail for both metallic and insulating systems.

[1] Gelb and Carnahan, Chem. Phys. Letts. 4-6 (2006) 283.

[2] Katayama et al., Nature 403 (2000) 170.

[3] Brazhkin et al., Physica B 265 (1999) 64.

[4] van Duin et al., J. Phys. Chem. A 105 (2001) 9396.

596f A Computational Study of Peptide Bond Hydrolysis

Bin Pan and Bernhardt L. Trout. Chemical Engineering, MIT, 77 Mass. Ave., Room E19-528, Cambridge, MA 02139

Protein molecules undergo various chemical degradation pathways, among which hydrolytic reaction of peptide bond is commonly observed. However, the detailed reaction pathway is poorly understood.

In this study, the hydrolysis reaction of a peptide bond under different solution conditions, namely, neutral and acidic pHs, was studied using a model compound N-methyl acetyl acrylamide (N-MAA). Ab initio molecular dynamics combining statistical mechanics and quantum mechanics was used to elucidate the reaction mechanism. Due to fuctuations at finite temperature in solution, path sampling method was used to generate an ensemble of trajectories according to their statistical weight in trajectory space. Car-Parrinello molecular dynamics was applied to advance the time evolution of the reaction in collecting these representative trajectories. Likelihood maximization was used to extract physically important degrees of freedom in the system and approximations of the reaction coordinate were compared. It was found that this hydrolysis reaction under both solution conditions is very complex in nature, which takes place in a concerted fashion and involves many degrees of freedom. PB histograms were computed to verify the results of reaction coordinate from likelihood maximization.

A coarse-grained model was developed to consider the surprisingly higher hydrolysis rate for the peptide bonds in the hinge region of antibodies molecules. Free energy for breaking the coarse-grained bond was calculated using umbrella sampling. The Bennett-Chandler procedure was also used to calculate reaction rate constants. It was found that the fuctuations in the system play an important role in determining the rate of breakage reaction. This may imply that the special structure of antibodies could cause the enhanced hydrolytic rate over peptides.

596g Ab Initio and Molecular Simulation of Viscoelastic Surfactants Solutions: Effect of Counter Ions

Abdullah Sultan[1], P.B. Balbuena[2], A.D. Hill[1], and H.a. Nasr-El-Din[1]. (1) Petroleum Engineering, Texas A&M University, 3116 TAMU - 714 Richardson Building, College Station, TX 77843, (2) Chemical Engineering, Texas A&M University, 3116 TAMU - 714 Richardson Building, College Station, TX 77843

Viscoelastic surfactants have a wide range of applications in the oilfield industry including: hydraulic fracturing and matrix acidizing. It has been proposed that surfactant molecules align themselves and aggregate due to intermolecular attraction or non-covalent bonds to form wormlike micelles as the ionic strength or pH increases ($pH > 2$). The micelles can further interact to form a network exhibiting elastic behavior. This increase in viscoelasticity helps to divert treating fluid while acidizing multiple zones with variation in permeability and it also has sufficient properties to suspend and carry the proppant into the factured zone.

Cationic and amphoteric surfactants have been extensively used to enhance well performance in carbonate geology. Several previous studies highlighted the significant role of counter ions in determining the apparent properties of aqueous solutions that contained viscoelastic surfactants. The effect of counter ions can be detrimental where the surfactant would not form a gel or precipitate, as in the case of ferric iron, in the formation and causes formation damage. Understanding how macroscopic properties depend on intermolecular interactions for complex fluids is a difficult task.

Molecular dynamics and molecular mechanics simulations are used to better understand how counter ions can impact physical properties of viscoelastic surfactants. Different molecular structures of cationic and amphoteric surfactants together with their counterions are simulated over a wide range of parameters. Simulation results such as density profiles, and radial distribution functions show a clear correlation between the surfactant head group and counterions. The results give important insight into the links between molecular structure and diverter (surfactant) phase behavior, which will help in developing more systematic procedures for treatment optimization, and the molecular design of more efficient/effective diverting systems.

596h Hydrazine Vapor-Liquid Equilibrium Prediction Using Gibbs Ensemble Monte Carlo Simulations with Opls-AA Force Field Parameters for Bonded Interactions Using Ab Initio Methods

Martha C. Mitchell, New Mexico State University, Dept. of Chem. Eng., Jett Hall 259, MSC 3805, Las Cruces, NM 88003-8001

Hydrazine is a highly reactive liquid base. It is mainly used as a rocket propellant; other uses include but are not limited to, the production of spandex fibers, agricultural chemicals, nickel-plating, nuclear fuel reprocessing, production of plastics and in the removal of halogens from wastewater. Molecular modeling of hydrazine can be a safe and effective method for predicting its chemical behavior during engineering applications including storage and transport, provided accurate molecular parameters are known. These simulations are essential for engineering design and safety analyses of any process that involves hydrazine. In this study ab initio calculations were used to determine the first OPLS-AA force field parameters for bonded interactions in hydrazine, which can be used in extensive molecular simulation studies. Gaussian 03 was used to determine equilibrium geometry data, bond force constants, angle force constants, and torsion parameters with geometry optimizations and single-point energy calculations using second-order Møller-Plesset perturbation theory (MP2), in the 6-31G(d,p) basis. The results are compared with published experimental data and are shown to be in good agreement. Using the updated parameters, Gibbs Ensemble Monte Carlo simulations were conducted to predict vapor-liquid equilibrium for hydrazine.

646a The Influence of Nickel Catalyst Geometry on the Dissociation Barriers of H2 and CH4: Ni13 Vs. Ni(111)

Bin Liu, Chemical Engineering, Colorado School of Mines, 1600 Illinois St., Golden, CO 80401, Mark T. Lusk, Physics, Colorado School of Mines, 1600 Illinois St., Golden, CO 80401, and James F. Ely, Chemical Engineering Department, Colorado School of Mines, 1500 Illinos St., Golden, CO 80401-1887.

The role of catalyst size and geometry in hydrocarbon decomposition is elucidated using density functional theory (DFT). For the sake of clarity, the work focuses on the dissociative adsorption of H_2 and CH_4 on an icosahedral Ni_{13} cluster and a Ni(111) surface. Both H_2 and CH_4 adsorb molecularly at t_1 sites of the cluster. Subsequent dissociation leaves H atoms most strongly bound to c_3 sites, while CH_3 favors t_1 and b_2 sites. A hybrid linear synchronous transit/quadratic synchronous transit (LST/QST) transition state search algorithm allows transition state structures to be located and dissociation barriers to be obtained. The lowest H_2 and CH_4 dissociation barriers are estimated to be 3.45 kcal/mol and 9.9 kcal/mol, respectively. The corresponding values for Ni(111) are 0.57 kcal/mol and 21.43 kcal/mol. Molecular orbital (MO) theory is employed to explain how local distortion in the electronic structure of the nickel atoms lowers these barriers. Meanwhile, a modified (111) surface is investigated to explore its effects on H_2, and CH_4 dissociation.

Using First-Principles DFT to Study the Reactivity of Hydrated Oxide Surfaces

Sara E. Mason and Anne M. Chaka. Physics, National Institute of Standards and Technology, 100 Bureau Dr., Gaithersburg MD 20899

Abstract

Geochemical reactions occurring at the interface between oxide surfaces and aqueous solution exert control on the transport and fate of contaminants. Molecular-level understanding of how metal ions adsorb to these surfaces is needed to understand contaminant speciation and bioavailability. Complexity arising from multiple surface stoichiometries, hydrogen bonding patterns, and varying oxygen functional group reactivities must be carefully considered. We use the interaction of lead with alumina and hematite as model systems to probe for essential factors governing the reactivity of hydrated surfaces towards contaminant ions. While geometrically isostructural, these two oxides have sharp contrasts in reactivity and electronic structure. We use this fact to make comparisons between the preferred adsorption modes of Pb(II) and discuss the relationships between surface structure, reactivity, and electronic structure.

Introduction

Lead (Pb) contamination in topsoil is one of the most serious environmental issues in the U.S. and other countries. Pb is toxic to almost all organs in the body, posing notable problems to neurological development, the hematological and cardiovascular systems, and the kidney. Despite its prevalence in the environment, molecular-level understanding of Pb adsorption is poorly understood.

The interfaces between natural soils and aqueous solutions make up a geochemically important part of the earth's surface where Pb ions adsorb and desorb, affecting the transport and ultimate fate of this contaminant.

In this work we present a detailed *ab initio* study aimed at understanding fundamental structure-reactivity relationships of Pb adsorption on model mineral-water interface systems. Fe- and Al-(oxyhyr)oxides comprise important examples of reactive mineral surfaces. Our study focuses on Pb(II) adsorbed to the hydroxylated (0001) surface termination of alumina and hematite. The fully hydroxylated, oxygen-terminated (0001) surface, denoted $(HO)_3$-M-M-*R* (where Fe or Al is indicated by M and *R* represents the continuing bulk atomic stacking sequence), is the dominant structure for hematite and the only one observed for alumina under geologically relevant conditions. By modeling Pb adsorption on the isostructural $(HO)_3$-M-M-*R* surface, we are able to isolate and study the reactivity-composition relationship.

Methodology

Periodic spin-polarized Density Functional Theory (DFT) calculations are carried out, using a double zeta, polarized atom-centered numerical basis set, as implemented in the Dmol3 code. Benchmarking and testing of our methods are carried out and compared to gas phase spectroscopy of the PbO molecular dimer and X-ray diffraction structures of crystalline PbO and PbO_2 oxides. We carry out additional benchmarking for other Pb compounds studied my various theoretical methods.

The results of our extensive benchmarking support the ability of our employed electronic structure methods to perform consistently and accurately across a range of possible Pb oxidation states and coordination environments.

The alumina and hematite surfaces are modeled by periodic slabs consisting of six O layers and 10 M layers. Periodic slab images are separated by no less than 25 Angstroms of vacuum. The two exposed surfaces are treated equivalently and related by inversion. We carry out a careful investigation of surface hydroxyl group orientation on both the hematite and alumina surfaces, and study this reactivity consideration. Full geometry optimizations are carried out with a structural convergence set to a force tolerance of 0.01 eV/Angstrom. To model Pb adsorption at geochemically feasible concentrations, large (2x2) surface supercells are used.

Initial Pb(II) adsorption sites are determined through consideration of Pb chemistry and the oxide surface geometry. In inorganic compounds, Pb is rarely in the elemental form and is generally present in the +2 oxidation state. Furthermore, it is resistant to redox reactions under geochemically relevant ranges in pH. The outermost O layer of the $(HO)_3$-M-M-*R* surface forms a triangular lattice. The centers of these O-triangles are inherently reasonable binding sites for Pb based on the experimentally observed trigonal pyramidal structure of Pb adsorption. The adsorbed structures are always charge-balanced such that Pb is in the naturally dominant +2 oxidation state. This requires displacement of 2 of the 3 H^+ at each adsorption site.

Applying electronic structure methods to environmental molecule science is an exciting and new area of research. The charge density, the essential quantity of DFT, can be analyzed in a number of ways to uncover important bonding interactions at reactive surfaces before and after adsorption. The electronic origins of Pb adsorption are studied using atom-projected density of states (PDOS) analysis. This semi-quantitative tool allows us to determine the atomic orbital character of the charge density. Covalent interactions are indicated in the PDOS by overlapping intensity between bonding atoms below the Fermi level, while ionic interactions are expressed by neighboring filled and empty valence states. Through PDOS analysis, we can identify the electronic interactions governing alumina and hematite surface reactivity and Pb binding.

Discussion

Our DFT comparative study of Pb(II) binding on isostructural hydrated alumina and hematite surfaces produces several trends that we study and explain through structural and electronic analysis. We construct a theoretical framework for Pb/water/oxide interactions with possible extension to other geochemically important interfaces.

Our periodic calculations model the surface H-bond arrangements as extended networks. We report two qualitatively different, but energetically similar, H-bonding patterns on both the alumina and hematite hydrated surfaces. The local geometry at the different Pb adsorption sites varies between the two reported energetically similar H-bond patterns. These two patterns therefore provide the opportunity to study the influence of hydroxyl group orientation on surface reactivity on both alumina and hematite. This is the first study addressing the impact of surface hydroxyl group orientation on hydrated surface reactivity.

We compute Pb adsorption energies that reproduce the experimentally established trend that hematite is more reactive towards Pb than alumina. What is unique to our result is that we determine the hematite surface to be more reactive based only on composition. The underlying origins of the varying surface reactivity can be traced to the partially occupied Fe *d*-band. Our analysis of the PDOS reveals two mechanisms through which the *d*-electrons influence the Pb binding. The two mechanisms are distinct in that one arises from the inherent hematite surface electronic structure, while the other is a localized interaction that depends strongly on the Pb binding geometry.

In the hematite surface, the Fe atoms are in a near octahedral environment. This coordination with surface oxygen atoms leads to a crystal field splitting in the Fe *d*-band. The spitting of the d-bands is very sensitive to the surface geometry. In the bulk, the Fe spins in hematite are in an anti-ferromagnetic configuration, alternating in sign between each Fe bilayer. Exposing the surface to either vacuum or water results in surface relaxations which result in varied Fe-O bond distances and therefore different Fe *d*-band energy splittings. We report how the hematite *d*-band is influenced by surface structure, and how this in turn influences electronic interactions with the Pb ion.

We also investigate the shape of the charge density for adsorbed Pb. We identify that the Pb lone pair is directed away from the oxide surface. By modeling additional molecular water, we demonstrate that the Pb lone pair is significantly hydrophobic. This result is likely general to Pb adsorbed on different (in both structure and composition) oxide surfaces. This result is key to assessing our methods in that it shows inappreciable stabilization provided by the addition of water molecules to the modeled systems.

Our results demonstrate the utility of DFT modeling to study and extract fundamental molecular-level understanding of geometrical interfaces. Our results span the steric effects imposed by hydroxyl group orientations, directionality of Pb charge, inherent alumina and hematite surface reactivity, hematite d-band splitting and it's sensitivity to surface structure, and Pb-surface electronic interactions.

Acknowledgments

This work was supported by NSF grants CBET-0404400 and CHE-0431425. Computational support and capabilities were provided by ARSC and NIH. S.E.M. was supported by an NRC-NIST Postdoctoral Fellowship.

References

G.E. Brown Jr., *et al.* Chem. Rev. **99**, 77 (1999).

G.E. Brown Jr., T.P. Trainor, and A.M. Chaka, in *Chemical Bonding at Surfaces and Interfaces*, edited by A. Nilsson. L. Pettersson, and J. Nørskov (Elesvier, Amsterdam, 2007), pp. 457-509.

T.P. Trainor, A.M. Chaka, P.J. Eng, M. Newville, G.A. Waychunas, J.G. Catalano, and G.E. Brown Jr., Surf. Sci. **573**, 204 (2004).

X.-G. Wang, A.M. Chaka, and M. Scheffler. J. Phys. Chem. **84**, 3650 (2000).

J.R. Bargar, G.E. Brown Jr., and G.A. Parks, Geochimica et Cosmochimica Acta **61**, 2617 (1997).

J.R. Bargar, G.E. Brown Jr., and G.A. Parks, Geochimica et Cosmochimica Acta **61**, 2639 (1997).

646c Classical Density Functionals for Modelling Hydrogen Physisorption at 77 K

Martin B. Sweatman, Chemical and Process Engineering, University of Strathclyde, 75, Montrose Street, Glasgow, G11XJ, United Kingdom

This work [1] surveys techniques based on classical density functionals for modelling the quantum dispersion of physisorbed molecular hydrogen at 77 K. Two such techniques are examined in detail. The first is based on the 'open ring approximation' (ORA) of Broukhno et.al. [2], and it is compared with a technique based on the semi-classical approximation of Feynman and Hibbs (FH). For both techniques, a classical density functional is used to model hydrogen molecule – hydrogen molecule (i.e. excess) interactions. Density functional results for a model of hydrogen adsorption into a graphitic slit pore at 77 K are compared with path-integral Monte-Carlo simulations. It is found that the ORA technique is flawed and that the best option is to combine the FH technique for hydrogen-surface interactions with an excess functional based on a calibrated hydrogen-hydrogen interaction.

[1] M.B. Sweatman, Physical Review E 77, 026712 (2008).

[2] A. Broukhno, P.N. Vorontsov-Velyaminov, and H. Bohr, Physical Review E 72, 046703 (2005).

646d First-Principles Theoretical Analysis of Dopant Adsorption and Diffusion on Surfaces of II-VI Compound Semiconductor Nanocrystals

Tejinder Singh, University of Massachusetts, Department of Chemical Engineering, Amherst, MA 01003, T. J. Mountziaris, Chemical Engineering, University of Massachusetts, Amherst, MA 01003, and Dimitrios Maroudas, Department of Chemical Engineering, University of Massachusetts, 686 N Pleasant Street, Amherst, MA 01003.

Nanocrystalline structures of compound semiconductors, such as the II-VI compounds ZnS, CdSe, and ZnSe, exhibit size-dependent optoelectronic properties and form the basis for a new generation of highly-integrated nanoelectronic and photovoltaic devices, as well as biological labels. Single-crystalline grains of these materials that are smaller than the corresponding exciton Bohr radius exhibit size-dependent luminescence due to quantum confinement. Doping in semiconductor nanocrystals could allow for precise control of their optical and electronic properties, thus enabling applications in solar cells and spintronics. However, in spite of its feasibility, doping of semiconductor nanocrystals has been an extremely difficult task. Self-purification of nanocrystals, an intrinsic mechanism of expulsion of impurities from the nanocrystal surface, makes the incorporation of dopants into nanocrystals even more difficult. In this context, we aim at obtaining a fundamental and quantitative understanding of dopant adsorption and diffusion on II-VI semiconductor nanocrystal surfaces, which can help elucidate the mechanisms of dopant incorporation into growing nanocrystals.

In this presentation, we report results on dopant adsorption and diffusion on surface facets of ZnSe nanocrystals based on first-principles density functional theory (DFT) calculations within the generalized gradient approximation. In our DFT calculations, we have employed slab supercells, plane-wave basis sets, and ultrasoft pseudopotentials. We have also implemented the nudged elastic band method including a climbing image to construct fully optimized dopant diffusion pathways and obtained accurate saddle-point configurations and activation energy barriers.

Specifically, we have focused on the incorporation of Mn dopant atoms into ZnSe nanocrystals through their surface facets. We found that the (001) facet is the dominant nanocrystal surface for Mn incorporation consistent with previously published findings [S. C. Erwin, *et al.*, Nature 436, 91 (2005)]. Our DFT calculations indicate that the binding energy for Mn adsorption onto various sites of the ZnSe (001)-(2x1) surface increases with increasing dopant surface concentration. This low binding energy at low dopant surface concentration provides an explanation for doping difficulties during nanocrystal growth. In addition, we have analyzed several dopant migration pathways for Mn diffusion on the ZnSe (001)-(2x1) surface. The calculated activation barriers for migration along the Se dimer rows range from 0.17 eV to 0.43 eV, depending upon the dopant surface concentration. Due to the low activation barriers, dopant atoms can migrate fast along the Se dimer rows without substantial surface relaxation. However, migration of the Mn atom across the Se dimer rows from the dimer site (Mn adsorbed onto the Se dimer) to the trough site (Mn adsorbed in the trough between Se dimers), is governed by a high-barrier pathway. At higher dopant surface concentrations, the binding energy for Mn adsorption onto the trough site is greater than that for Mn adsorption onto the dimer site. In conjunction with the high activation barriers for Mn migration in the trough parallel to the Se dimer rows, this implies that migration to a trough site may lead to dopant incorporation into the ZnSe nanocrystal

The Nature of Proton Conduction in a Polymer Electrolyte Membrane, Nafion

Yoong-Kee Choe[1], Eiji Tsuchida[1], Tamio Ikeshoji[1], Shunsuke Yamakawa[2], Shi-aki Hyodo[2], [1]National Institute of Advanced Industrial Science and Technology (AIST), Tsukuba Japan. [2]Toyota Central R&D Labs., Inc., Nagakute, Japan.

Introduction

Proton transport in the aqueous environment has attracted much attention for more than a century. It is well known that protons are transported by the Grötthuss mechanism (proposed two hundred years ago), which is sometimes referred to by terms such as proton jumping and proton hopping. However, the proton transport mechanisms in constrained systems such as biological membranes, electrolyte membranes and minerals are rather vague, despite the fact that proton transport plays an important role in the functionality of such systems. The focus of study in this paper is a polymer electrolyte membrane, which is placed between the anode and cathode in polymer electrolyte fuel cells (PEFCs). Detailed knowledge of the proton transport mechanisms in these membranes is extremely important because the generation of power by fuel cells is limited by proton conduction. The nature of the proton conduction in such a constrained environment is also very interesting from a purely scientific point of view. Nafion is a membrane widely used in PEFCs due to its very good proton conductivity. Several mechanisms of proton transport have previously been proposed in the literature, but have not yet yielded precise descriptions on the atomistic level; various details of these mechanisms are still unclear and controversial. We believe that this lack of consensus regarding the proton conduction mechanisms results mainly from a lack of direct and unambiguous information provided by experimental studies. However, such an investigation may be more suited to highly sophisticated molecular simulations. Therefore, in the present paper, we have applied first-principles molecular dynamics simulations to investigate the nature of the proton conduction in Nafion.

Computational Details

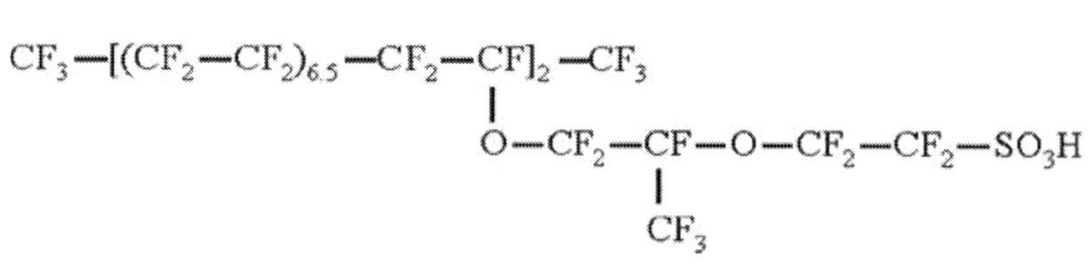

Figure 1. Schematic illustration of modeled Nafion

To obtain reasonable initial configurations for the use of the first-principles MD simulations, we started by carrying out classical MD calculations. Nafion was modeled as a polymer consisting of two monomers, as described in Fig. 1. Because a computation of the entire Nafion structure including cylindrical channels of water clusters with the first-principles MD is not possible, we examined the basic sandwich structure consisting of the water region with hydrophilic side chains and surrounding hydrophobic regions. The MD simulations were carried out in a periodic three-dimensional cell, in which Nafion and water were stacked in the z-direction. Models with two different water contents were examined. The lattice constants of the x- and

y-axes were fixed in order to represent the surface area of the micelle that was occupied by each sulfonic acid end group. The lattice constant of the z-axis was determined in order to represent the composite density of dry Nafion and water at a temperature of 363 K. The molecular numbers and the lattice constants of unit cells in the models are listed in Table 1. In the present study, two humidity levels, water content (λ = number of water molecules/number of sulfonic acid groups) equal to 4.1 and 12.7, were considered and hereafter we call the former the *L-state* (*low water content state*) and the latter the *H-state* (*high water content state*). Fujitsu Materials Explorer 3.0 was used for this simulation, which was performed assuming the *NVT* ensemble at 363 K using the Nosé–Hoover thermostat. The total simulation time was approximately 200 ps with a time step of 0.2 fs. The proton that originated from the sulfonic acid end group was treated as a hydrated ion (H_3O^+). H_2O and H_3O^+ molecules were treated using the extended simple point charge (SPCE) water model. The Dreiding force field was adopted for all of the intramolecular potential forces of Nafion. Universal force field (UFF) parameters were used to model the nonbonded interactions between H_2O (H_3O^+) and Nafion. The point charge present on each atom was determined by the electrostatic potential (ESP) method after molecular orbital calculations. Snapshots of the simulated final configurations of Nafion/water complexes are shown in Fig. 2. The hydrophilic side chains were oriented toward the water region, and thus, the two-phase structure was maintained. Then, using the configurations obtained from the classical MD simulation as an initial configuration, first-principles MD simulations were performed. The electronic structures of the systems were evaluated by means of density functional theory in the Kohn–Sham formalism. We used the generalized gradient approximation in the Perdew–Burke–Ernzerhof (PBE) form, which is expected to give an accurate description of the hydrogen bonds. Separable norm-conserving pseudopotentials were employed, and only the Γ point was used to sample the Brillouin zone. Adaptive finite elements were used in place of ordinary plane waves for the basis set. All production runs in the present paper were performed with an average cutoff energy of 57 Ry, while the resolution was approximately doubled at the positions of fluorine and oxygen atoms by adaptation of the grids. In the molecular dynamics part, the nuclei were treated classically, and their equations of motion were integrated using the velocity–Verlet algorithm with the forces calculated from the electronic structure. The temperature was controlled by the Berendsen thermostat with a target temperature of 353 K. All hydrogen atoms in the system were given the mass of deuterium to allow the use of a large time step of 1.21 fs in the production run. However, the deuterium atoms are denoted by "H" in the following analysis whenever there is no possibility of confusion. During the molecular dynamics simulations, the electronic states were quenched to the Born–Oppenheimer surface at every time step with the

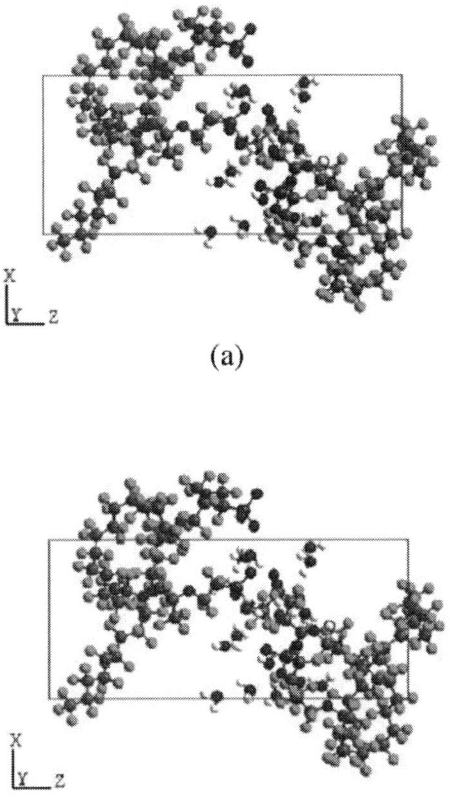

Figure 2. Snapshots of the configurations of (a) *L-state* and (b) *H-state*.

limited-memory variant of the quasi-Newton method. The wave functions were extrapolated from previous time steps. First, equilibration runs up to 10 ps were carried out without an electric field and further simulations without an electric field were carried out up to 20 ps. Then, we performed simulations considering an electric field up to 10 ps. All of the calculations were carried out with our own finite element DFT code *FEMTECK* (Finite Element Method-based Total Energy Calculation Kit). The above method has been successfully applied to liquid molecular systems. The electric field was evaluated using the Berry phase theory, which is most appropriate for treating disordered systems under periodic boundary conditions. We applied an electric field of 0.5 V/nm along the x-direction of the cells to achieve acceptable statistical accuracy at reasonable computational cost, while avoiding noticeable changes in liquid structures. In the present study, because of the limitation of the model, we mainly focus on proton dynamics occurring within a few Å (~8 Å) from $-SO_3^-$. Proton transport from one hydrophilic pocket to another one is beyond the target of the present investigation.

Table 1 Summary of simulation conditions

	Condition	
	L-state	*H-state*
Molecule numbers in a unit cell		
N_{Nafion}	2	2
N_{H_2O}	13	47
$N_{H_3O^+}$	4	4
Water content λ		
$(N_{H_2O} + N_{H_3O^+})/N_{SO_3^-}$	4.25	12.75
Size of unit cell		
x/Å	12.3042	12.3042
y/Å	12.3042	12.3042
z/Å	28.1239	35.1549

Results

At first, we investigate the size and the connectivity of water clusters. The results of cluster analysis shown in Fig. 3 indicate, the hydrogen bond network in the *L-state* is not continuous, forming small water clusters. In contrast, the network is connected to a large extent in the *H-state*, as in bulk water. This feature is in agreement with the result of a classical MD study which employed much larger model systems than ours.

According to the analysis of the trajectories, in agreement with experimental observations, the diffusion coefficient of protons in the *L-state* is calculated to be much smaller than that in the *H-state.* To investigate the origin of the slow proton diffusion in the *L-state* in more detail, we analyzed the hopping probability of the protons. We obtained hopping probabilities of 5.6% for the *H-state* and 4.5% for the

L-state. During 100 simulation time-steps, this implies that proton hopping occurs on average 5.6 times in the *H-state* and 4.7 times in the *L-state*. The difference between the hopping probabilities in the two states is small, a result that is at odds with the general concept of the vehicular mechanism. Thus, the question arises as to why the proton diffusion in the *L-state* is much slower. An answer can be found by examining the nature of the proton hopping in the *L-state*. As described above, the hydrogen bond network of water in the *L-state* is broken to some extent. Therefore, Grötthuss-like structural diffusion is permitted only within the water clusters. Thus, the water molecules in the *L-state* behave analogously to those in supercritical conditions. The disconnected hydrogen bond network of water in the *L-state* causes proton transfer to occur in a *non-constructive* manner; a proton is transferred temporarily from a water molecule to its neighboring molecule by hopping, but then returns to its original location (Details see Fig. 4). In contrast, as seen in Fig. 4, we can also define proton transfer that contributes constructive to proton conduction. Many proton transfers within small water clusters in the *L-state* can be described as non-constructive, and these processes do not contribute significantly to the actual diffusion of protons. The accurate microscopic description of proton motion revealed from the simulations is accordingly completely different to that of the vehicular mechanism, which assumes protons transport as bonded to a "*vehicle*" water. The extraordinarily high mobility of protons in an aqueous environment has been explained by the concept that proton hopping facilitates proton conduction significantly. However, our computations reveal, in contrast to the generally accepted concept, that the occurrence of proton hopping alone does not guarantee good proton mobility. What is equally important is the structure of water hydrogen bond network, which limits paths that protons can travel around. Poor and good conductivities of protons inside Nafion, therefore, can be reasonably explained by the ratio of the constructive and non-constructive proton transfer events. An increase of the occurrence of the former relative to the latter increases the mobility of protons.

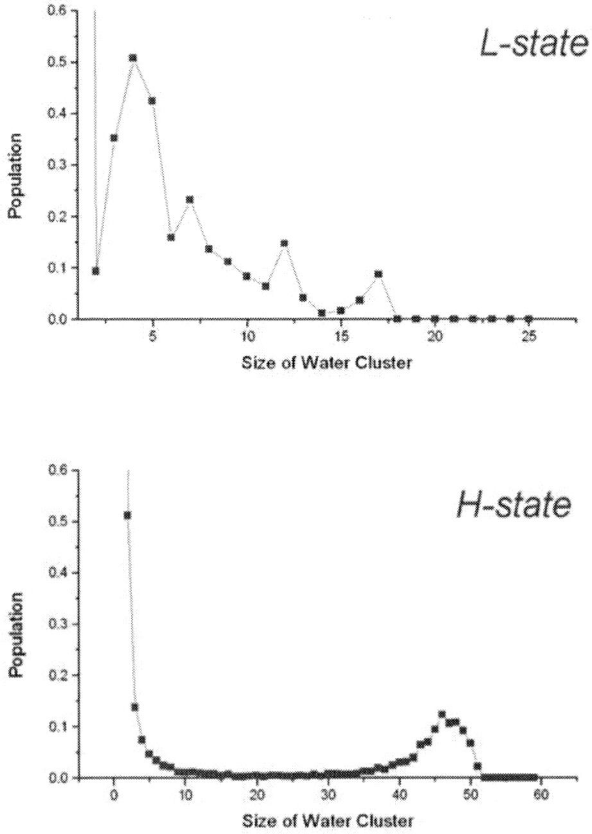

Figure 3. Results of the cluster analysis. A hydrogen bond was said to b e present if the O···H bond length was less than 2.0 Angstrom. In this analysis, we identified clust ers of water molecules connected by a hydrogen bond network in all snapshots. The number of wat er molecules in each cluster was then extracted and the average population of each size of cl uster was calculated. Populations of clusters with a variety of sizes were found in the *L-state*. Conversely, the population of clusters composed of 10 - 40 water molecules in the unit cell was nearly zero in the *H-state*, which indicates that water molecules in the H-state have very similar characte r to those in bulk water.

More details of our results such as proton dynamics occurring next to sulfonic group will be presented at the meeting. Here we summarize main conclusions derived from our findings. we have used first-principles MD simulations to investigate the nature of proton conduction in Nafion. The hydrogen bond network of water is completely different in the *L-state* and *H-state*, being discontinuous in the former and continuous in the latter. Analysis of the trajectories for each state shows that the proton hopping probability has little dependence on water content. For the *L-state*, our simulations show that the diffusion coefficients of the protons and water are very similar, which is in accord with previously reported experimental observations. However, the results of our simulations demonstrate that the previously proposed explanation for the slow diffusion in terms of a simple vehicular-like mechanism does not correctly describe the motion of protons on an atomistic level. Indeed, proton hopping between adjacent water molecules occurs in both the *L-state* and *H-state*, but due to the disconnected nature of the water hydrogen bond network in the *L-state*, this type of proton hopping does not contribute constructively to the proton conduction.

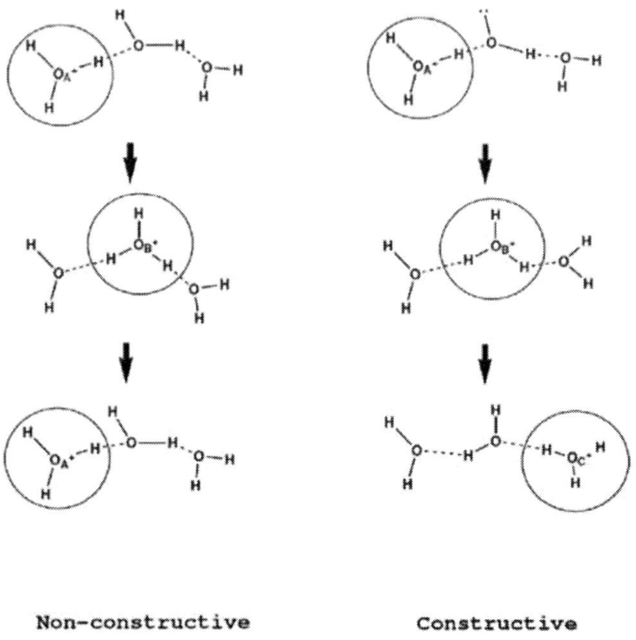

Figure 4. The constructive proton transfer is the transfer of a proton t hat moves further away from its original location. The non-constructive proton transfer is the trans fer of a proton that comes to the position of the oxygen site of the original water in the end. Contributi on to proton conduction is larger in the former and smaller in the latter.

References

1. Cui, S.; Liu, J.; Selvan, M. E.; Keffer, D. J.; Edwards, B. J.; Steele, W. V. *J. Phys. Chem. B* **2007**, *111*, 2208.
2. Tsuchida, E. *J. Chem. Phys.* **2004**, *121*, 4740.
3. Tsuchida, E. *J. Phys. Soc. Jpn.* **2006**, *75*, 054801
4. Choe, Y.-K.; Tsuchida, E.; Ikeshoji, T. *J. Chem. Phys.* **2007**, *126*, 154510.
5. Choe, Y.-K.; Tsuchida, E.; Ikeshoji, T. *Comp. Phys. Comm.* **2007**, *177*, 38.
6. Choe, Y.-K.; Tsuchida, E.; Ikeshoji, T., Yamakawa, S.; Hyodo S. *J. Phys. Chem. B* **2008**, *112*, 11586.

646f Elemental Mercury Adsorption Study on Gehlenite ($Ca_2Al_2SiO_7$)

Bo Gyeong Kim and Paul Blowers. Chemical and Environmental Engineering, The University of Arizona, PO Box 210011, Tucson, AZ 85721-0011

The pure components (CaO, $Al_2O_3 \cdot SiO_2$, $CaCO_3$ and etc.) of PWDS (Paper Waste Derived Sorbent) rearrange to gehlenite ($Ca_2Al_2SiO_7$) during the mercury adsorption process at 600 – 1100 °C. The rearranged gehlenite surface captures mercury on the active sites and then is deactivated. At high temperatures compared to mercury adsorption on activated carbon, this sorbent captures 83 – 90 % of elemental mercury according to bench-scale Hg control experiments. The mercury removal efficiency increases with increasing temperature. In this work, computational chemistry is used to investigate how mercury is adsorbed on the gehlenite surface and to theoretically predict the temperature affects in an effort to mechanistically understand the unusual adsorption behavior.

The adsorption of Hg on the gehlenite surface was investigated with density functional theory (DFT). The gehlenite structure was modeled with a periodic slab and was optimized with the local density approximation (LDA) with the Perdew – Wang (PWC) functional. The adsorption energies of Hg adsorbed on the gehlenite surface was calculated with the gradient corrected (GGA) method with the Becke-Lee-Yang-Parr (BLYP) functional using the lower level optimized structure. The fine basis set, core treatment and spin unrestricted methods were applied to all calculations to lead higher level and hopefully more accurate results. The equilibrium constants with respect to the temperature effects were also predicted, the results from this work show how Hg adsorbs on an atop site of oxygen with a shorter distance and stronger adsorption energy on gehlenite compared to other adsorbates.

646g Understanding the Silver Nanoparticle Growth Via First Principle Methods

Giannis Mpourmpakis, Center for Catalytic Science and Technology, Department of Chemical Engineering, University of Delaware, Colburn lab, 150 Academy Street, Newark, DE 19716 and Dion Vlachos, Director of Center for Catalytic Science and Technology (CCST), University of Delaware, Newark, DE 19716.

First principle calculations have been performed to examine the growth mechanisms of silver nanoparticles (NPs) in the presence of citrate. Through suitable energy relations, we show that the carboxyl groups preferentially interact with specific crystallographic planes of NPs and solvent effects can be significant. As the charge of the NP approaches zero, electrostatic interactions gradually decrease, and this affects binding and the role of citrate. Further reduction of the citrate-silver NP complex decreases the binding energy of the citrate. This decrease is related to the symmetry of the NP via the latter's electron affinity value. We found out that hydrogen bonds among citrate molecules play a key-step in the growth mechanism. Our results explain for the first time the effect of pH and citrate concentration on the morphology of silver NPs and provide evidence that it is possible to unravel and control NP size and shape.

669a Efficient Potential of Mean Force Calculations through Monte Carlo Simulations

Lorant Janosi and **Manolis Doxastakis**. Chemical and Biomolecular Engineering, University of Houston, S222 Engineering Bldg. 1, Houston, TX 77204

Potential of mean force (PMF) calculations along a reaction coordinate require exhaustive sampling which imposes severe computational limitations for any algorithm applied. The Expanded Ensemble Density of States method (EXEDOS) constitutes an elegant formalism to sample continuously and uniformly the reaction coordinate and extract the desired PMF. While the algorithm is trivial to parallelize, the iterative estimation of the required weights can be rendered exceedingly slow through insufficient sampling of the remaining degrees of freedom. In the present study, we propose a combination of Monte Carlo methods that can overcome the computational barriers present in the algorithm in a variety of systems where the reaction coordinate is the separation distance between two molecules. We applied the EXEDOS formalism in a simple well-studied system of two large colloids immersed in a solution of small particles. We demonstrate that calculations at all densities can be significantly accelerated through selective sampling of translational degrees of freedom in the vicinity of the large colloids. At low densities, application of rejection free Monte Carlo moves for the large particles can result in further improvement by almost an order of magnitude in computational time. At high densities where simple Monte Carlo simulations are prohibitively demanding and rejection-free moves fail, we explore recently proposed algorithms for reversible mapping between local minima that could provide a unique route for accurate PMF calculations in dense molecular systems.

669b Rigorous Algorithms for Ensemble Overlap and Model Development Using the Relative Entropy

M. Scott Shell, Chemical Engineering, University of California, Mail Code 5080, Santa Barbara, CA 93106

The inverse problem in statistical physics asks: given an observed set of properties, what are the molecules, energy functions, and/or thermodynamic conditions that give rise to them? Such a task is at the heart of multiscale strategies, in which one desires a coarse-grained or low resolution model that faithfully reproduces the properties of a detailed one. There have been several specific approaches to this problem, including the reverse Monte Carlo method [Lyubartsev and Laaksonen, Phys. Rev. E 52, 3730 (1995)] and the force-matching approach [Ercolessi and Adams, Europhys. Lett. 26, 583 (1994); Voth and Izvekov, J. Phys. Chem. B 109, 2469 (2005)]. However, it has been challenging to identify general theoretical approaches, for arbitrary systems, energy functions, and ensembles.

Here we show that the relative entropy, Srel = sum p_T ln(p_T/ p_M), provides a fundamental overlap metric and unifying framework for inverse molecular-thermodynamic problems, involving optimization of a model system (`M') to reproduce the properties of a target one (`T'). We demonstrate that the relative entropy serves as a generating function for principles in variational mean field theory and uniqueness, and gives intuitive results for simple case scenarios in model development. Importantly, we show that the relative entropy provides new numerical techniques for linking models at different resolutions, by coupling its determination with flat-histogram and transition-matrix-based Monte Carlo algorithms. We also show that the relative entropy carries physical significance by using it to quantify the deviations of a three-site model of water from simple liquids; importantly, we demonstrate a surprising close connection between the relative entropy and kinetic anomalies in water. We then use the relative entropy to find optimal simple and conformal models that represent complex fluids and mixtures.

669c A Fast Heuristic for Predicting Ordered Arrangements of Strongly Interacting Particles

Eric Jankowski, Chemical Engineering, University of Michigan, Ann Arbor, MI 48109 and Sharon C. Glotzer, Chemical Engineering and Materials Science and Engineering, University of Michigan, 2300 Hayward Street, Ann Arbor, MI 48109.

Advances in synthetic chemistry, lithography, and microfabrication continue to expand the library of shapes and materials that can be used to synthesize new nanoparticles. By combining a hierarchy of interaction energies with complex geometries to form novel building blocks we aim to enable the self-assembly of more complicated materials and devices, but it also becomes more difficult to predict what structures these new building blocks will form. With a range of complex anisotropic interactions available to these particles, systems can become easily trapped in metastable states during Metropolis Monte Carlo (MC) simulations, making it computationally difficult if not impossible to find their energy minimizing configurations even with cluster-based MC methods. We present a new computational method - binary hierarchical assembly - that rapidly generates zero temperature, energy-minimizing configurations of sets of complex nanoparticles. Our new algorithm generates lower energy structures than can be predicted with standard cluster Monte Carlo techniques in a fraction of the time and facilitates an understanding of how small changes in nanoparticle detail can translate to large changes in macrostructure. We provide a quantitative measure to determine if a given system is a good candidate for binary hierarchical assembly. We discuss why and under what conditions our new method and other optimization techniques are more efficient than Monte Carlo methods for strongly interacting systems of particles.

669d Development and Application of a Hybrid Method Involving Interpolation and *Ab Initio* Calculations for the Determination of Transition States

Anthony Goodrow[1], Alexis T. Bell[1], and Martin Head-Gordon[2]. (1) Department of Chemical Engineering, University of California at Berkeley, 201 Gilman Hall, Berkeley, CA 94720-1462, (2) Department of Chemistry, University of California at Berkeley, 419 Latimer Hall, Berkeley, CA 94720-1460

Determination of transition states (TS) is important for understanding the kinetics of chemical reactions. Methods such as the nudged-elastic band (NEB) require an intial guess for the geometry of the TS and electronic structure calculations can fail if the guess is not correct. By contrast, the growing-string method (GSM) requires only reactant and product geometries in order to initiate the search for the TS; however this method is computationally intensive because a large number of gradient calculations need to be made. Several modifications have been implemented in order to reduce the computational cost of the GSM. First, internal coordinates have replaced Cartesian coordinates because they are better able to describe the movement of atom(s) during elementary reactions. Second, the conjugate gradient method has replaced the steepest descent method during the minimization of orthogonal forces in the search of points along the minimum energy path (MEP). Third, an interpolation scheme has been implemented to estimate the energy and gradient rather than using quantum mechanics (QM) code, thereby saving significant computational time. The interpolation scheme that was implemented was adopted from work previously done on reaction dynamics [M. A. Collins, Theor. Chem. Acc. **2002**, *108*, 313]. The modified GSM has been tested on four cases of increasing complexity: the Müller-Brown potential energy surface, alanine dipeptide isomerization, H-abstraction during methanol oxidation on isolated vanadate species supported on silica [A. Goodrow, A. T. Bell, J. Phys. Chem. C **2007**, *111*, 14753], and C-H bond activation in the oxidative carbonylation of toluene to p-toluic acid by Rh complexes [X. Zheng, A. T. Bell, J. Phys. Chem. C **2008**, *112*, 2129]. The modified GSM represents a 2-3 time reduction in computational effort over the previous method as measured by the number of QM gradients computed, without a sacrifice in the accuracy of the geometry and energy of the final TS. Additional savings in computational effort can be achieved by carrying out the initial search for the MEP using a lower level of theory, e.g. HF/STO-3G, and then refining the MEP using DFT at the B3LYP level with larger basis sets.

669e Multicomponent Gauge Cell Monte Carlo Method: Determination of Chemical Potentials in Dense and Inhomogeneous Systems

Aleksey Vishnyakov, Deparment of Chemical and Biochemical Engineering, Rutgers University, 98 Brett Road, Piscataway, NJ 08854-8058 and Alexander V. Neimark, Department of Chemical and Biochemical Engineering, Rutgers University, 98 Brett Road, Piscataway, NJ 08854-8058.

We present a modification of the ideal gas gauge cell (IGGC) Monte Carlo simulation method (A. V. Neimark and A. Vishnyakov, J. Chem. Phys. 122, 234108 (2005)) designed for chemical potential calculations in dense and inhomogeneous multicomponent systems. To measure the chemical potentials of individual components of an m-component system, the simulation is carried out in m+1 simulations cells: the system under study (system cell) is set in chemical equilibrium with m gauge cells, one gauge cell per component. The system cell and the gauge cells are immersed into the thermal bath of a given temperature. Each gauge cell represents a finite volume reservoir, which contains only one component of the system. The molecules transferred from the system cell to the gauge cell change their identity and behave as ideal particles. The size of the gauge cell controls the level of density fluctuations for the given component in the system. This scheme, called the mesoscopic canonical ensemble, bridges the gap between the canonical and the grand canonical ensembles, which are known to be inconsistent in small systems. The chemical potential of a system component is defined as the work of insertion of one additional molecule, and it is rigorously calculated from the multidimensional equilibrium distribution of particles between the system and gauge cells. It is shown that the IGGC method is more efficient than the Widom method, especially in dense and inhomogeneous small systems. The proposed method is verified in the case of a binary mixture of supercritical methane and subcritical pentane, which features a liquid to vapor transition as the methane concentration increases. A special attention is paid to extreme confinements of several molecular diameters in the cross-section, where the inconsistency between the canonical ensemble and the grand canonical ensemble is most pronounced. The multicomponent IGGC method is illustrated drawing on the example of mixture of strongly and weakly adsorbing components confined to spherical pores with attractive walls.

Web Page: sol.rutgers.edu/aneimark.html

669f Including Convection In Lattice Kinetic Monte Carlo Simulations

Matthew H. Flamm, Scott L. Diamond, and Talid R. Sinno. Chemical and Biomolecular Engineering, University of Pennsylvania, 311A Towne Building, 220 South 33rd Street, Philadelphia, PA 19104

Biological processes such as platelet aggregation in a blood vessel and biofilm formation are complex, multiscale phenomena that are influenced by fluid flow [1,2]. While such phenomena can be modeled entirely on the basis of continuum descriptions, the large number of discrete particles that typically must be considered suggests the application of mesoscopic simulation frameworks, such as Dissipative Particle Dynamics [3], kinetic Monte Carlo [4], and others.

In this talk, we present a lattice kinetic Monte Carlo (LKMC) model for simulating the trajectories of multiple diffusing particles that are subjected to an externally applied flow field, i.e. the flow field is not affected by the mesoscopic distribution of particles, although it can be time varying in general. The LKMC approach effectively coarse-grains out the details of single particle morphology, and is ideally suited for the study of aggregation processes because of its computational efficiency. However, it is generally applied to purely diffusive situations.

The primary input to LKMC simulations is a rate database for each possible particle move on the lattice. We first demonstrate that accounting for the local flow field around a particle with a simple bias in its rate in the direction of the flow leads to a large error in the particle trajectories, even for the case of simple one-dimensional flow. Through a detailed mathematical analysis, we show the origin of this error, and present a new algorithm that allows for the correct simulation of particle motion in systems where both diffusion and convection are present. The algorithm and analysis first are validated by comparing LKMC predictions to analytical results for Taylor-Aris dispersion in parallel plate flow. The new LKMC model is then applied to more complex situations that include two-dimensional flows. Detailed comparisons are made to continuum numerical simulations.

[1] I. V. Pivkin, P. D. Richardson, G. Karniadakis, Proceedings of the National Academy of Sciences 103, 17164 (2006)

[2] C. Picioreanu, M. C. M. van Loosdrecht, J. J. Heijnen, Biotechnology & Bioengineering 72, 205 (2001)

[3] R. D. Groot, P. B. Warren, The Journal of Chemical Physics 107, 4423 (1997)

[4] K. A. Fichthorn, W. H. Weinberg, The Journal of Chemical Physics 95, 1090 (1991)

669g New Techniques for Simulating Crystals

Martin B. Sweatman, Chemical and Process Engineering, University of Strathclyde, 75, Montrose Street, Glasgow, G11XJ, United Kingdom

The crystalline state is fundamental in nature and hugely important in modern technology. Yet, despite this, molecular simulation techniques for classical crystals at equilibrium are, overall, not as satisfactory as those for fluids, and in particular there is some confusion in the literature as to how confined crystals can be simulated correctly. This presentation will describe some recent work that has made some progress in this area. First, the Gibbs ensemble simulation technique applied to solid – fluid phase coexistence [1] will be introduced. Then, a new 'self-referential' technique [2,3] able to calculate crystal free energies straightforwardly and efficiently, and which should allow easier simulation of confined crystals, will be described.

1. M.B. Sweatman and N. Quirke, 'Simulating fluid - solid equilibrium with the Gibbs ensemble', Mol. Sim. 30, 23 (2004).

2. M.B. Sweatman, 'A self-referential Monte-Carlo method for calculating the free energy of crystalline solids', Phys. Rev. E. 72, 016711 (2005).

3. M.B. Sweatman, A.A. Atamas and J.M. Leyssale, 'The self-referential method combined with thermodynamic integration', J. Chem. Phys. 128, 064102 (2008).

709a International Study of R&D In Simulation-Based Engineering and Science: Overview and Executive Summary

Sharon C. Glotzer, Chemical Engineering and Materials Science and Engineering, University of Michigan, 2300 Hayward Street, Ann Arbor, MI 48109

We present an overview of a recently conducted international assessment of R&D in Simulation-Based Engineering and Science. Key findings are presented and discussed. The study was sponsored by NSF, DOE, NIH, NIST, NASA, and DOD. Panelists include: Peter Cummings, Abhi Deshmukh, Sharon Glotzer (study chair), Martin Head-Gordon, George Karniadakis, Sangtae Kim (study co-chair), Linda Petzold, Celeste Sagui, and M. Shinozuka.

709b Simulation-Based Engineering and Science: Materials Applications and Multiscale Modeling and Simulation for Materials, Energy and Life Sciences

Peter T. Cummings, Department of Chemical Engineering, Vanderbilt University, Nashville, TN 37235-1604

Text Not Available.

709c An International Assessment of R&D In Simulation-Based Engineering and Science: Introduction

Phillip R. Westmoreland, Division of Chemical, Bioengineering, Environmental and Transport Systems, National Science Foundation, Arlington, VA 22230

Text Not Available.

709d Simulation-Based Engineering and Science - Discussion

Phillip R. Westmoreland, Division of Chemical, Bioengineering, Environmental and Transport Systems, National Science Foundation, Arlington, VA 22230

Text Not Available.

709e Simulation-Based Engineering and Science: Applications to Energy, and Big Data & Data-Driven Simulations

Sangtae Kim, Chemical Engineering, Purdue University, 585 Purdue Mall, West Lafayette, IN 47907

Text Not Available.

709f Simulation-Based Engineering and Science: Applications to Life Sciences & Medicine, and Next-Generation Hardware & Software

Linda R. Petzold, University of California, Department of Computer Science, Santa Barbara, CA 93106

Text Not Available.

734a Thermodynamic Investigation of the System Oxygen + Carbon Dioxide + Cyclohexane + Cyclohexanol + Cyclohexanone

Thorsten Merker[1], Jadran Vrabec[2], and Hans Hasse[1]. (1) Laboratory of Engineering Thermodynamics, University of Kaiserslautern, Erwin-Schroedinger Str. 44, Kaiserslautern, 67663, Germany, (2) Institute of Thermodynamics and Thermal Process Engineering, University of Stuttgart, Pfaffenwaldring 9, Stuttgart, 70569, Germany

To develop novel routes for the catalytic oxidation of cyclohexane, using carbon dioxide expanded media, it is aimed to explore and describe the thermodynamics of mixtures of oxygen, carbon dioxide, cyclohexane, cyclohexanol, and cyclohexanone. Experimental work is combined with molecular modeling and simulation. The key properties are vapor-liquid equilibria (VLE) and diffusivities.

In a first step, the thermodynamics of the bulk phase are studied. An equation of state based model for phase equilibria, thermal and caloric properties of the studied system is provided. The present work focuses on molecular modeling and simulation of the relevant pure substances and their mixtures. The molecular models are of the multi-center Lennard-Jones plus quadrupole, dipole or point charge type. Reliable molecular models for oxygen [1] and cyclohexane [2] are available from prior work, models for the other components are being developed here. The Grand Equilibrium method [3] is used for the calculation of the vapor-liquid equilibria. The expanded ensemble method [4] is applied for calculating the chemical potential in the liquid phase for large, strongly interacting molecules. New molecular models for carbon dioxide, cyclohexanol, and cyclohexanone [5] are presented. The mean unsigned errors in the full temperature range from triple point to critical point are for carbon dioxide 0.8 % in saturated liquid density, 0.9 % in vapor pressure, and 5 % in heat of vaporization. The analogous numbers for cyclohexanol are 0.4 % in saturated liquid density, and 4.1 % in vapor pressure. These models are also capable to predict transport properties properly. Therefore, the self- and transport diffusion coefficients are determined by equilibrium molecular dynamics simulation with the Green-Kubo method [6]. VLE for the investigated mixtures are determined and compared to experimental data where possible.

References

[1] Vrabec, J.; Stoll, J.; Hasse, H. J. Phys. Chem. B. 2001, 105, 12126-12133.

[2] Eckl, B.; Vrabec, J.; Hasse, H. J. Phys. Chem. B. 2008, submitted.

[3] Vrabec, J.; Hasse, H. Molec. Phys. 2002, 100, 3375-3383.

[4] Vrabec, J.; Kettler, M.; Hasse, H. Chem. Phys. Letters. 2002, 356, 431-436.

[5] Merker, T.; Vrabec, J.; Hasse, H. in preparation

[6] Fernandez, G. A.; Vrabec, J. ; Hasse, H. Int. J. Thermophysics. 2004, 25, 175-186.

SYSTEMATIC STUDY ON THE SOLUBILITY OF GASES IN ALKYL-IMIDAZOLIUM BASED IONIC LIQUIDS FROM A MOLECULAR MODELING APPROACH

Jordi S. Andreu, Institut de Ciència de Materials de Barcelona, Consejo Superior de Investigaciones Científicas (ICMAB-CSIC) and MatGas Research Center, Bellaterra. Barcelona, Spain.
Lourdes F. Vega, Institut de Ciència de Materials de Barcelona, Consejo Superior de Investigaciones Científicas (ICMAB-CSIC), MatGas Research Center and Air Products and Chemicals Inc, Bellaterra. Barcelona, Spain.

Introduction

Much scientific and technological interest is currently focused on room temperature ionic liquids (ILs). This is due the "designer" properties of some ILs, particularly those based on the imidazolium cation where the physical properties of the liquid may be controlled by incorporation of appropriate functional groups, along with the environmentally friendly properties of ILs, most notably negligible vapor pressure, high thermal stability, good solvent ability toward a wide range of solutes, etc., making ILs attractive candidates as replacements for volatile organic solvents. A careful characterization of them is needed prior they are put into final use for a specific application; this includes knowledge of the relationship between the structure and the properties. This knowledge can be obtained by experimental techniques, simulations and/or theoretical approaches. Theoretical approaches offer the advantage of the speed in which these calculations are performed, provided they are accurate enough to reproduce the experimental data.

The soft-SAFT Equation of State

Within the framework of SAFT, the EoS of a fluid is a perturbation expansion given in terms of the residual molar Helmholtz energy. The Helmholtz energy is defined as the difference between the total Helmholtz energy and that of an ideal gas at the same temperature T and molar density ρ. SAFT implicitly assumes that there are three major contributions to the total intermolecular potential of a given molecule: *a)* the repulsion-dispersion contribution typical of individual segments *b)* the contribution due the fact that these segments can form long lived chains and *c)* the contribution due to the possibility that some segments form association complexes with other molecules. Furthermore, we added a new term a^{polar} which takes into account the quadrupolar interactions between molecules, in case they exist. According to this scheme, the residual Helmholtz energy can be written as:

$$a^{res} \equiv a - a^{ideal} = a^{ref} + a^{chain} + a^{assoc} + a^{polar} \qquad (1)$$

Here, a^{res} is the residual Helmholtz free energy density of the system. The superscripts *ref, chain, assoc* and *polar* refer to the contributions from the reference term, the formation of the chain, the association, and the polar interactions, respectively, depending on the system under study.

In the soft-SAFT EoS, [1], [2], [3] the reference term is a Lennard-Jones (LJ) spherical fluid, which accounts for both the repulsive and attractive interactions of the monomers forming the chain. As in previous works the accurate EoS of Johnson et al. [4] is used here. For the case of mixtures the same equation is used by applying the van der Waals one-fluid theory, with generalized Lorentz-Berthelot mixing rules:

$$\sigma_{ij} = \eta_{ij}\left(\frac{\sigma_{ii}+\sigma_{jj}}{2}\right) \tag{2}$$

$$\varepsilon_{ij} = \xi_{ij}\left(\varepsilon_{ii}\varepsilon_{jj}\right)^{1/2} \tag{3}$$

here η and ξ are the size and energy binary adjustable parameters, respectively. The equation is used in a purely predictive manner from the pure component parameters when η and ξ are equal to unity, while values different from unity mean the use of one or two binary parameters, taking into account the differences in size and/or energy of the segments forming the two compounds in the mixture.

The chain and association terms come from Wertheim's theory [5], and they are formally identical in the different versions of SAFT:

$$a^{chain} = \rho k_B T \sum_i x_i (1-m_i)\ln g_{LJ} \tag{4}$$

$$a^{assoc} = \rho k_B T \sum_i x_i \sum_\alpha \left(\ln X_i^\alpha - \frac{X_i^\alpha}{2}\right) + \frac{M_i}{2} \tag{5}$$

here ρ is the molecular density, T is the temperature, m is the chain length, x_i is the molar fraction of component i, k_B the Boltzmann constant and g_{LJ} is the radial distribution function of a fluid of LJ spheres at density $\rho_m = m\rho$, evaluated at the bond length σ. M_i the number of associating sites of component i, and X_i^α the mole fraction of molecules of component i non bonded at site α, which accounts for the contributions of all the associating sites in each specie.

The leading multipolar term for fluids of linear symmetrical molecules, like carbon dioxide, nitrogen, acetylene, etc., is the quadrupole-quadrupole potential. An expansion of the Helmholtz free energy density in terms of the perturbed quadrupole-quadrupole potential with the Padé approximation was proposed by Stell et al. [6]:

$$a^{qq} = a_2^{qq}\left(1 - \frac{a_3^{qq}}{a_2^{qq}}\right)^{-1} \tag{6}$$

Expressions for a_2 and a_3, the second and third-order perturbation terms, were derived for an arbitrary intermolecular reference potential [7], [8] and involve the state variables, molecular parameters, and the integral for the reference fluid. A detailed derivation of these expressions is given elsewhere [9]. This new term in the soft-SAFT EoS involves an additional molecular parameter, Q, the quadrupolar moment, which is usually taken from experimental data.

The molecular model for gases and the ionic liquids

A needed step before applying the equation to mixtures is to obtain the molecular parameters of the pure compounds. We have used the molecular parameters for CO_2, H_2 and Xe obtained in previous works [10], [11], without any further fitting. In this way, the transferability of these models as well as the performance of the equation is enhanced. We refer to those previous works for a detailed explanation on those models.

Two different models for the ILs have been used in this work. The first model is applied for the alkyl-imidazolium ILs with the $[BF_4]^-$ and $[PF_6]^-$ anions [12]. Based on results obtained from molecular dynamics simulations and experimental evidences [13], [14], [15], we have modeled these ILs as LJ chains enclosing anion plus cation together and using one associating site in each molecule, which accounts for the competition between counter ions from different molecules. The chain length, size and energy parameters were obtained by fitting the equation to experimental density data available in literature; the specific data was selected according to its robustness and wide thermodynamic range. In order to keep the minimum number of fitted parameters, at first approximation, the association parameters previously used for the alkanols family were used, thus avoiding further fitting.

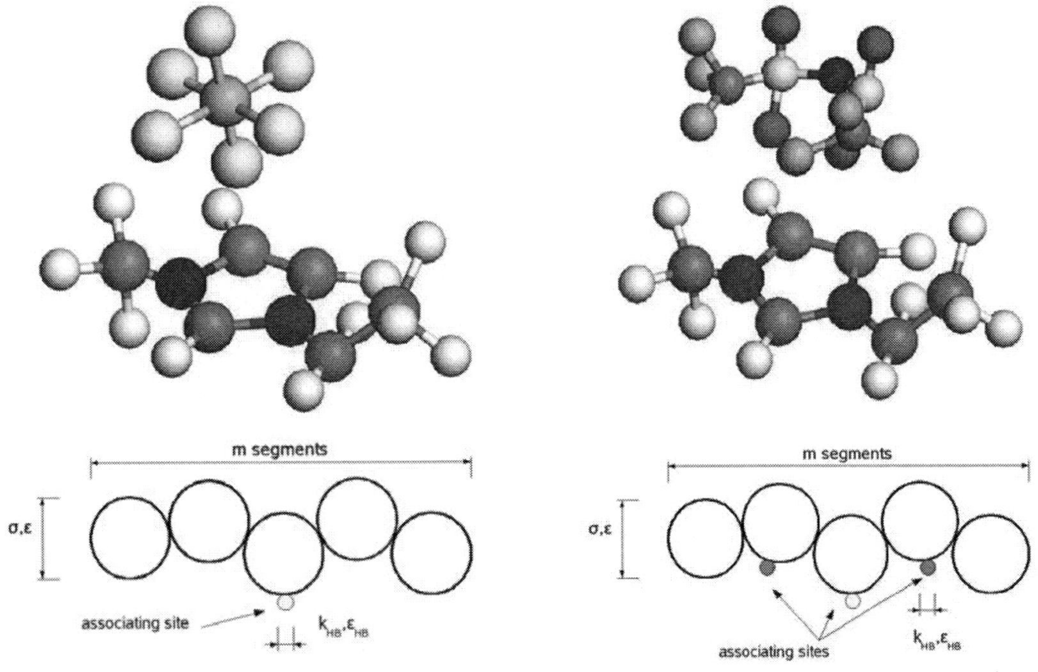

Figure 1. Sketch of the models used to describe the alkyl-imidazolium families of ILs studied in this work. The associating sites account for the competitive attraction between counterions. Left: alkyl-imidazolium [PF6] ILs, right alkyl-imidazolium [Tf2N] ILs.

The members of the alkyl-imidazolium $[Tf_2N]$ family have been modeled with a different approach [16]. Since in the soft-SAFT EoS an associating site is used to represent short range and highly directional attractive forces between molecules, we used this feature to mimic the main interactions between counter ions. The main interaction between counterions is due the coulombic forces as a consequence of their opposite charge. Moreover, the delocalization of the anion electric charge due the oxygen atoms attached to the nitrogen enhance the possibility of interaction with the surrounding cations through them. According to these interactions we used a site *A* representing the nitrogen atom and 2 *B* sites representing the delocalized charge due to oxygen atoms on the anion. Each type of associating site is identically defined, but only *AB* interactions were allowed between different IL molecules. As well as for the model used for the other two ILs presented in this work, the chain length, size and energy parameters were fitted to experimental density data while the the same associating parameters values were also used for this family. This was done for two different reasons: to avoid further fitting of parameters and to check the transferability of the new model. A sketch of these two models is provided in figure 1.

Results I: Pure compounds

Following previous works, we were able to correlate the molecular parameters of the ILs with their molecular weight. Three of the molecular parameters for the alkyl-imidazolium family correlate with the molecular weight, while the other two parameters (those corresponding to the association) were kept constant for the whole family of ILs. None of these parameters are temperature dependent and this implies that those models are independent of the thermodynamic conditions. The correlation for the molecular parameters of the $[C_n\text{-mim}][PF_6]$ is as follows:

$$m = 0.0188M_w - 0.7590 \quad (7)$$
$$m\sigma^3\ (Å^3) = 1.926M_w - 221.350 \quad (8)$$
$$m\varepsilon/k_B\ (K) = 8.540M_w - 514.820 \quad (9)$$

The correlation for the molecular parameters of the $[C_n\text{-mim}][BF_4]$ is:

$$m = 0.0188M_w + 0.2440 \quad (10)$$
$$m\sigma^3\ (Å^3) = 1.847\ M_w - 119.760 \quad (11)$$
$$m\varepsilon/k_B\ (K) = 8.530\ M_w - 40.123 \quad (12)$$

and the correlation for the molecular parameters of the $[C_n\text{-mim}][Tf_2N]$ reads as:

$$m = 0.0059M_w + 3.4919 \quad (13)$$
$$m\sigma^3\ (Å^3) = 1.877\ M_w - 313.670 \quad (14)$$
$$m\varepsilon/k_B\ (K) = 2.898\ M_w - 1000.700 \quad (15)$$

These correlations enable the equation with predictive power, as they provide the possibility to predict the behavior of heavier members of these series, not included in the fitting procedure, providing then the possibility of studying compounds for which few solubility data have been reported. Moreover, this is particularly useful for the case of ILs, as it has been demonstrated than the toxicity of the ILs increases as the alkyl chain length increases, making experimental measurements more dangerous [15].

Once the models for the pure compounds were developed and parameters obtained, they are used to predict the solubility behavior of hydrogen, carbon dioxide and xenon in the selected imidazolium families of ILs. Results concerning the application of these models to available data are presented and discussed in the next section.

Results II: Solubility of gases in alkyl-imidazolium ionic liquids

We present and discuss here calculations performed with the aforementioned models with the soft-SAFT EoS as applied to three families of ILs with H_2, CO_2 and Xe, compared with available experimental data.

We first studied the mixtures [C_4-mim][BF_4] + CO_2 and [C_4-mim][PF_6] + CO_2 at low pressures in a wide range of temperature (T=283K-348K). The soft-SAFT calculations obtained were done with pure component parameters, and hence they are pure predictions. The accuracy of these calculations as compared to the experimental data is good, indicating that the simple model presented in the previous section is able to capture the solubility of these mixtures in quantitative agreement with experimental data. However, small deviations in the [PF_6] family were found and the use of one independent temperature parameter (ξ=0

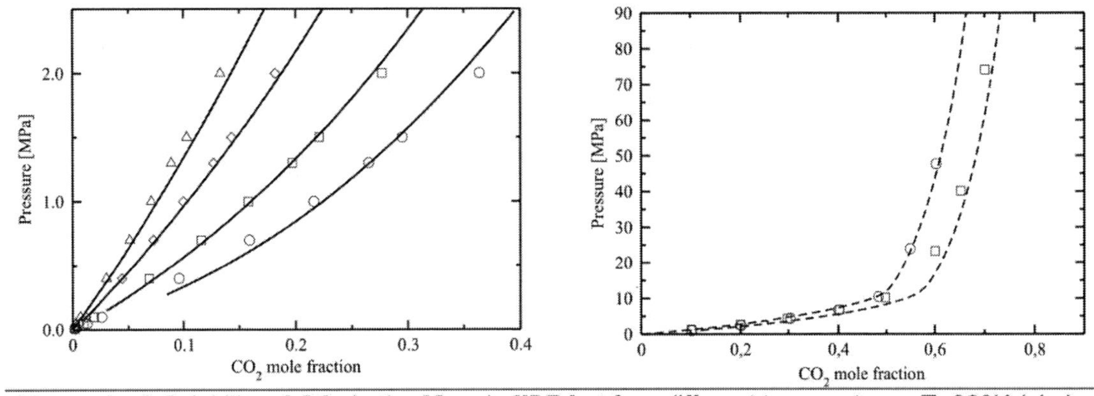

Figure 2. a) Solubility of CO_2 in the [C_4-mim][BF_4] at four different temperatures: T=283K (circles), T=298K (squares), T=323K (diamonds) and T=348K (triangles). Solid lines are predictions with soft-SAFT model without fitting. **b)** Solubility of CO2 in the [C_n-mim][BF_4] at 330K. Symbols represent experimental data for n=4 (circles) and n=6 (squares). Dashed lines corresponds to the soft-SAFT model with ξ=0.985 at n = 2, 4.

Predictions at high pressures and for other members of these families of ILs were performed next (figure 3). The model is able to capture the sharp increase in pressure at a given CO_2 composition, without any fitting. However, quantitative predictions with experimental data were not obtained in this case. Soft-SAFT is able to quantitatively reproduce the behavior of these mixtures at high pressures with a single binary temperature independent parameter ξ=0.985 (very close to unity) for all members of the [PF_6] family.

This transferable parameter allows the model to capture the experimental trend exhibited by these systems even at high concentrations of CO_2, reproducing the sharp rise in the pressure with the correct slope in almost quantitative agreement with experimental data, as in the previous case. Note that the ξ parameter (equation 3) takes into account the differences in energy of the segments forming the chains (van der Waals type interactions), no s

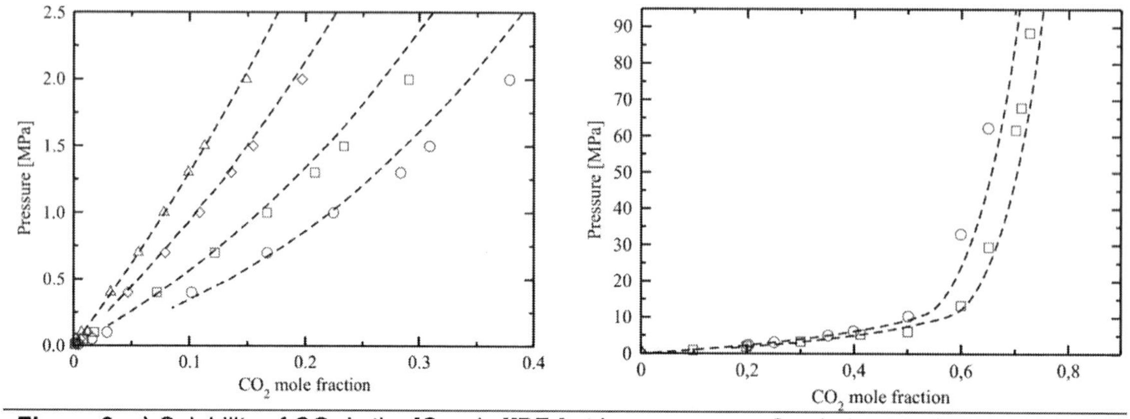

Figure 3. a) Solubility of CO_2 in the [C_n-mim][PF_6] at low pressures. Symbols as in figure 2a. In that case, the use of a binary parameter independent of temperature is needed to fit the experimental data (ξ=0.970). **b)** Solubility of CO_2 in [C_4-mim][PF_6] and [C_6-mim][PF_6] at high pressures. Symbols are experimental data for n=4 (circles) and n=6 (squares). In this case, the same binary parameter value is used at low and high pressures.

Next, results concerning the applicability of the soft-SAFT model developed for the Tf_2N family are presented. Pure predictions for the systems [C_n-mim][Tf_2N] + CO_2 were used to test the model of the pure compounds. As in the previous cases with simpler ILs, the main trends of these systems were well described without any fitting of the binary parameters. However, a temperature independent mixture parameter was needed to describe the experimental data available for these systems, obtaining excellent results. In that case, a different value of this parameter is needed for each member of the ionic liquid family (ξ=0.985 for n=2; ξ=0.972 for =4; ξ=0.948 for n=6), contrary to the previous case in which one parameter sufficed for the whole family. This situation can be understood through the more complex nature of the anion, which has a more complex atomic structure and conformation.

According to the good performance of these models describing the solubility behavior of H_2 and Xe in these ILs, and also to the small values obtained for energy binary parameter, the equation was used as a predictive tool to compare the different solubility behavior of these three gases in these families of alkyl-imidazolium based ILs, at similar thermodynamic conditions.

Two comparative predictions of the solubility of gases in alkyl-imidazolium ILs using the aforementioned models are presented next. First, we compare the solubility of CO_2 and H_2 in different members of the [C_n-mim][Tf_2N] family at T=333K and in [C_n-mim][PF_6] family at T=330K (figure 4). We observe that the highest absorption capability of these gases is found for the ILs with the longest alkyl chain length. The same behavior was found for mixture with xenon.

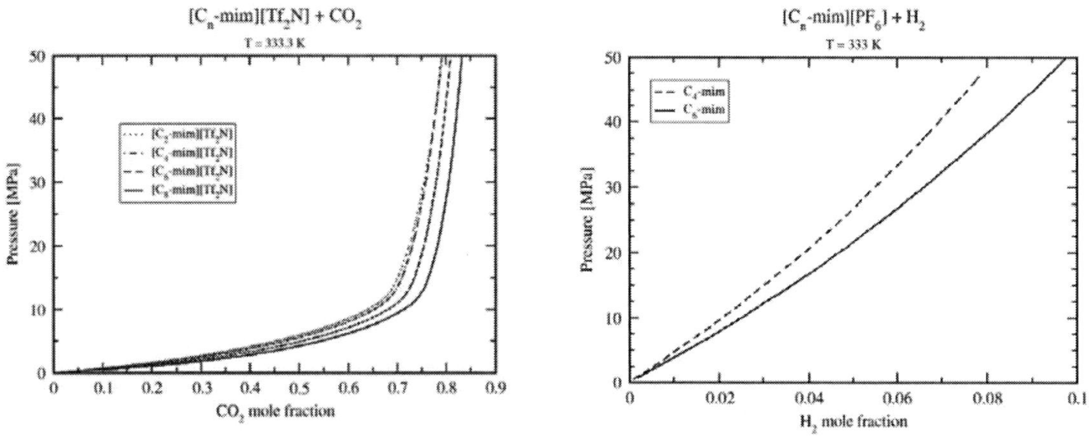

Figure 4. Solubility of carbon dioxide in [C_n-mim][Tf_2N] for n=2 (dotted line), n=4 (dashed-dotted line), n=6 (dashed line) and n=8 (solid line) at T=333.3K using the optimized parameters. We see that CO_2 solubility increases as the chain length of the imidazolium ring increases (left). The same behavior is observed for mixtures with hydrogen and the alkyl imidazolium hexafluorophosphate IL (right).

Secondly, we compare the strong anion dependence on the absorption of H_2, CO_2 and Xe (figure 5), showing that the [Tf_2N]$^-$ anion enhances the absorption within the imidazolium family.

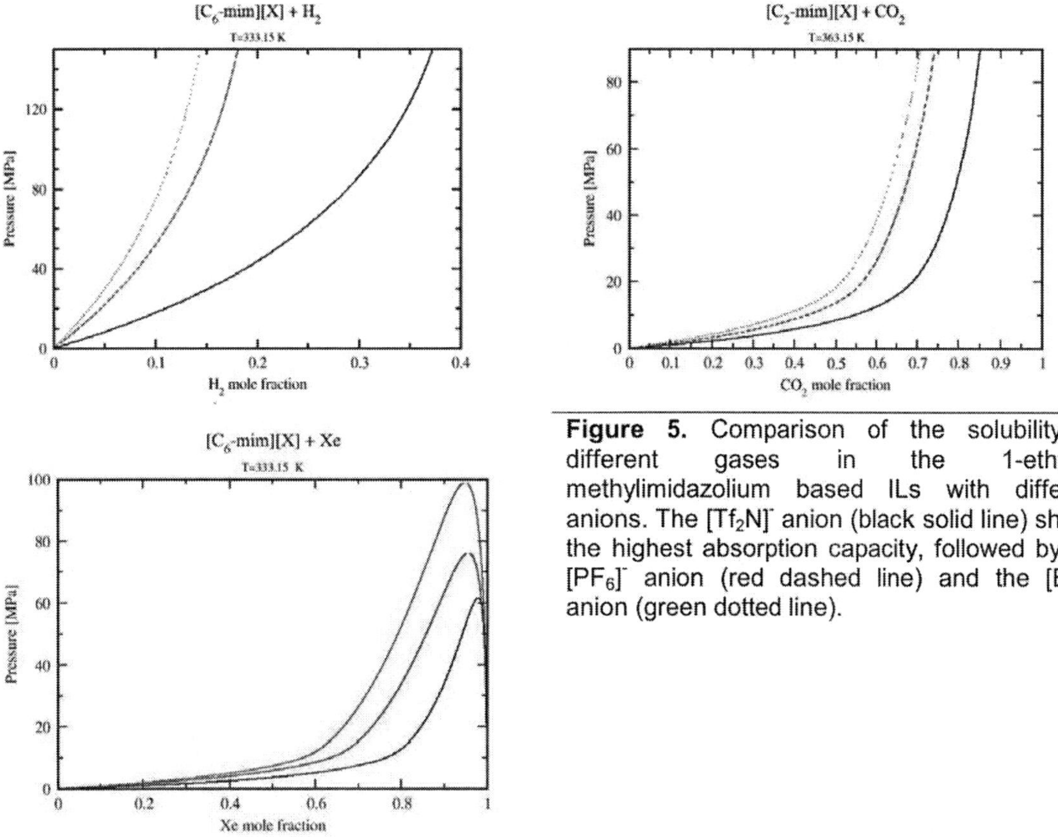

Figure 5. Comparison of the solubility of different gases in the 1-ethyl-3-methylimidazolium based ILs with different anions. The [Tf_2N]$^-$ anion (black solid line) shows the highest absorption capacity, followed by the [PF_6]$^-$ anion (red dashed line) and the [BF_4]$^-$ anion (green dotted line).

Conclusions

We have shown the capability of soft-SAFT to reproduce the solubility of H_2, CO_2 and Xe in three different families of ILs in a wide range of temperatures and pressures with a simple model for the IL. The key of the model is to mimic the strong pair interaction between the anion and the cation as a specific associating site in the IL molecule. Only five parameters were needed to model the ILs while the association parameters were kept constant throughout the three families. The models were able to capture the sharp rise in pressure as the solubility of the gases increases from pure component parameters, without the need of any specific assumption regarding the interaction of the ILs with the gases. Moreover, the robustness of the soft-SAFT equation as well as the models developed provides a fast tool which can be used to evaluate the absorption capacity of this alkyl-imidazolium ILs for others gases of interest.

Acknowledgements

Discussions with Drs. Brian K. Peterson and Carlos Rey-Castro are gratefully acknowledged. Financial support has been provided by the Spanish Government, Ministerios de Educaciốn y Ciencia y Ciencia e Innovaciốn, (projects CTQ2005-00296/PPQ and CTQ2008-05370/PPQ) and the Catalan Government (2005SGR-00288). J.S.A. acknowledges a grant from MATGAS 2000 AIE.

References

1. F.J. Blas, L.F. Vega (1997), *Mol. Phys.*, 92, pp. 135.
2. J.C. Pàmies, L.F. Vega (2001), *Ind. Eng. Chem. Res.*, 40, pp. 2532.
3. F.J. Blas, L.F. Vega (1998), *Ind. Eng. Chem. Res.*, 37, pp. 660.
4. J.K. Johnson, J. Zollweg, K.E. Gubbins (1993), *Mol. Phys.*, 78, pp. 591.
5. M.S. Wertheim (1984-1986), *J. Stat. Phys.*, 35, 19, Ibid., 42, pp. 459, 477.
6. G. Stell, J.C. Rasaiah, H. Narang (1974), *Mol. Phys.*, 27, pp. 1393.
7. C.H. Two, K.E. Gubbins, C.G. Gray (1975), *Mol. Phys*, 29, pp. 713.
8. C.H. Two, (1976), *PhD Dissertation, University of Florida.*
9. K.E. Gubbins, C.H. Two (1978), *Chem. Eng, Sci.*, 33, pp. 863.
10. A.M.A. Dias, H. Carrier, J.L. Daridon, J.C. Pàmies, L.F. Vega, J.A.P. Coutinho, I.M. Marrucho (2006), *Ind. Eng. Chem. Res.*, 45, pp. 2341.
11. J.C. Pàmies, L.F. Vega (2001), *Ind. Eng. Chem. Res.*, 40, pp. 2532.
12. J.S. Andreu, L.F. Vega (2007), *J. Phys. Chem. C*, 111, pp. 16028.
13. S.R. Urahata, M.C.C. Ribeiro (2004), *J. Chem. Phys*, 120 (4), pp. 1855.
14. T.I. Morrow, E.J. Maginn (2002), *J. Chem. Phys. B.*, 106, pp. 12807.
15. M.G. Del Pópolo, G.A. Voth (2004), *J. Phys. Chem. B.*, 108, pp. 1744.
16. J.S. Andreu, L.F. Vega (2008), *J. Phys. Chem.* (in press).
17. J. Ranke, K. Mölter, F. Stock, U. Bottin-Weber, J. Poczobutt, J. Hoffmann, B. Ondruschka, J. Filser, B. Jastorff (2004), *Ecotoxicol. Environ. Saf*, 58, pp. 396.

734c Computer Simulation of Hybrid Organic-Inorganic Nanocomposites

Maurizio Fermeglia[1], Sabrina Pricl[2], Paola Posocco[3], Marek Mály[4], and Martin Lisal[4]. (1) Molecular Simulation Engineering Laboratory, Department of Chemical Engineering, University of Trieste, Piazzale Europa 1, Trieste, 34127, Italy, (2) Molecular SImulation Engineering (MOSE) Laboratory - DICAMP, University of Trieste, Piazzale Europa 1, Trieste, 34127, Italy, (3) University of Trieste, Piazzale Europa 1, Trieste, 34127, Italy, (4) E. Hala Laboratory of Thermodynamics, Institute of Chemical Process Fundamentals, Academy of Sciences of the Czech Republic, 6-Suchdol, Prague, 165 02, Czech Republic

A current challenge of physical, chemical and engineering sciences is to develop theoretical tools for predicting structure and physical properties of hybrid organic inorganic nanocomposite from the knowledge of a few input parameters. However, despite all efforts, progress in the prediction of macroscopic physical properties from structure has been slow. Major difficulties relate to the fact that (a) the microstructural elements in multiphase material are not shaped or oriented as in the idealizations of computer simulations, and more than one type can coexist; (b) multiple length and time scales are generally involved and must be taken into account, when overall thermodynamic and mechanical properties wish to be determined, and finally (c) the effect of the interphases/interfaces on the physical properties is often not well understood and characterized. As a consequence, their role is often neglected in the development of new theoretical tools or they are treated in a very empirical way. In this work, we focused on issues (b) and (c) in a multiscale molecular simulation framework, with the ultimate goal of developing a computationally-based nanocomposite designing tool.

In this work a computational procedurewas developed, which allowed to obtain atomistic models of hybrid organic/inorganic networks based on 3-glycidyloxypropyltrimethoxysilane (GPTMS), an organofunctional alkoxysilane monomer that can undergo both the sol-gel polymerization of the alkoxy groups and curing of the epoxy functionality to form an I/O hybrid network with covalent bonds between organic and inorganic phases. The actual procedure, however, could be applied with minor modifications to any similar system. The simulation allowed also to estimate several thermophysical properties of the 3D network, such as mechanical moduli and constant pressure specific heat, whose values were found to be in the range of the corresponding experimental evidence for similar systems. These results confirm both the accuracy of the 3D model structure generated by the script procedure, and the quality of the force field used. Therefore, the proposed computational recipe can constitute a useful tool for the design and development of new hybrid I/O systems with improved structural/chemico-physical properties.

734d Study on Gas Combustions In a Downhole Steam Generator by Reactive Dynamics Simulations

Luzheng Zhang[1], Robert Lee[1], and Adri Van Duin[2]. (1) Petroleum Recovery Research Center, New Mexico Institute of Mining and Technology, 801 Leroy Place, Socorro, NM 87801, (2) Materials and Process Simulation Center, 139-74, California Institute of Technology, 1200 E. California Blvd., Pasadena, CA 91125

There are very large heavy oil reserves in many countries, US, Canada, etc, however, conventional primary recovery techniques are unapplicable because the high viscosity of these oils, often in excess of 1000 centipoises, prevents them from flowing freely to the surface. Various other recovery methods are used instead, the application of which is mostly depth limited. For deeper reservoirs, steam injection from the surface is used generally, but its effectiveness is limited in most cases to about a 2500 ft depth duo to the losss of heat. To overcome this limitation, various attempts have been made to locate the steam generator downhole (downhole steam generator, DSG). Comparing to a surface steam generator, the DSG are needed to opearte under different conditions, resulting in a special requirements on fuel and oxidizer for the combustion. Thus, an optimal fuel and its oxidizer has to be selected for better oil recovery in a specific resevoir condition.

To address fundamental issues (energy release, temperature increase, produced gases, etc), we experimented gas combustions of various combinations of fuels (methanol, ethanol and propane) and oxidizers (oxygen, ozone and hydrogen peroxide) at different conditions (temperature, pressure, fule/oxidizer ratio, etc.) by molecular dynamics (MD) simulations with the reactive force field (ReaxFF). System energy, temperature, pressure, produced gases and their composition as well as their overall kinetics are analysized in detail for all stuided systems. In practice, major handicaps in the DSG include the excessive rate of metal corrosion by the hot combustion products in the presence of wet steam and the possible incomplete combustion because of the limitted well diameter. A catalystic combustion may be a good approach to eliminating these problems in the DSG. Thus, we also simulated the catalytic reactions in gas comubstion for a few systems to test the possibility how a catalyst can improve the combustion in DSG. Finally, we suggest a combination of fuel and its oxidizer for the design of the DSG for a better oil recovery.

734e The Effects of Water Structure on the Kinetics and Thermodynamics of Xylan Hydrolysis

Haitao Dong[1], Mark Nimlos[2], and Xianghong Qian[1]. (1) Department of Mechanical Engineering, Colorado State University, Fort Collins, CO 80523-1374, (2) National Bioenergy Center, National Renewable Energy Laboratory, 1617 Cole Blvd., M/S 3322, Golden, CO 80401

One of the critical issues in biochemical conversion of lignocellulosic biomass to biofuels is sugar yields. Understanding the kinetics and thermodynamics of the depolymerization process and the sugar loss reactions including degradation and condensation reactions during dilute acid pretreatment is essential to optimize the pretreatment conditions. Car-Parrinello based ab initio molecular dynamics (CPMD) simulations with advanced metadynamics (MTD) technique are used to investigate the reaction mechanisms, the reaction free energies, barriers and rate constants. Protonation is found to be the rate limiting step in both hydrolysis and sugar degradation/condensation reactions. The proton affinity of water structure surrounding the reactants will impact the reaction barriers significantly. In particular, the role of hydrogen bonding interaction on these reactions will be elucidated.

734f Mechanism of Spontaneous Initiation in High-Temperature Polymerization of N-Butyl Acrylate: a Theoretical Study

Sriraj Srinivasan[1], Myung Won Lee[2], Michael C. Grady[3], Masoud Soroush[1], and Andrew M. Rappe[4]. (1) Department of Chemical and Biological Engineering, Drexel University, 32nd & Chestnut Streets, Philadelphia, PA 19104, (2) Department of Chemistry, University of Pennsylvania, 231 South 34th Street, Philadelphia, PA 19104, (3) DuPont Marshall Lab, 3401 Grays Ferry Ave., Philadelphia, PA 19146, (4) Makineni Theoretical Laboratories, Department of Chemistry, University of Pennsylvania, 231 South 34th Street, Philadelphia, PA 19104-6323

The growing requirement for environmental friendly polymeric resins has rendered a major change in the manufacturing of paints and coatings. Low-solids, high-molecular weight acrylic resins have been replaced with highly functional, high-solids, low-molecular-weight resins [1, 2, 3]. Such desired resins have been produced mainly using high temperature (ca. above 100 oC) polymerization. The dynamics of high-temperature polymerization are very different from its low-temperature counterpart, as at high temperatures different reaction mechanisms govern the process [4, 5]. For this reason, low-temperature polymerization models have been found to be inherently incapable of predicting the dynamics of high-temperature polymerization. Efforts have been made to understand better the high-temperature polymerization dynamics. Grady et al. [6] discovered sustained, reproducible, spontaneously-initiated polymerization of n-butyl acrylate at high temperatures in the absence of any known extraneously added initiators. Quan et al. [7] characterized the polymers using electronspray ionization/Fourier transform mass spectrometry and nuclear magnetic resonance spectroscopy, to understand chain branching reactions. Felix et al. [8] performed kinetic studies using mechanistic macroscopic modeling to evaluate contribution of different reactions to overall rate of polymerization. So far, these laboratory and marcroscopic-modeling studies have been inconclusive in identifying the initiation mechanism and the initiating species in spontaneous thermal polymerization of acryl acrylates.

Spontaneously initiated thermal polymerization has been known for styrene and methyl methacrylate [9]. Mayo's mechanism of self initiation is widely recognized as the mechanism of initiation in styrene and other cyclic molecules [10,11]. Diradical mechanism of self-initiation proposed by Flory [12] is speculated to initiate high temperature polymerization of methyl methacrylate (MMA), although no experimental evidence is available as of yet.

In this study, we present results from density functional theory calculations [13] that we have conducted to identify the initiating species and the mechanism of initiation in high-temperature thermal polymerization of n-butyl acrylate. The study involves calculation of reactant, transition state, intermediate, and product geometries using B3LYP/6-31G(d). The molecular geometries of n-butyl acrylate monomer and diradical have been calculated on the singlet and triplet potential energy surfaces [14]. The rate constants and activation energy of the diradical formation have been calculated. The diradical has been found to be a stable intermediate on the triplet energy surface, which is in agreement with previously known hypotheses [9]. The presence of a stable Diels-Alder (DA) intermediate, a dimer, has been found on the singlet energy surface, which had never been reported before. The calculated rate constant for diradical formation was found to be few orders of magnitude lower than the previously reported experimental values [8]. This difference can be due to errors arising from experimental measurements that had been made in time-scales (in minutes) higher than those known for formation of diradicals (10-6–10-9 s). Monoradicals are widely recognized to initiate polymerization reactions, not diradicals. In view of this, we have studied the formation of monoradicals via hydrogen abstraction by the diradical and the DA dimer from a third monomer. Activation energy and rate constant for the monoradical formation have been calculated, and the resulting values have been compared to the literature values reported in [8].

References

[1] Schultz, K.R., Proc-Adv. Coat. Technol. Con, Chicago, IL, 1992.

[2] VOC's Directive EU Committee of the American Chamber of Congress in Belgium, ASBL/VZw, Brussels, July 8, 1996.

[3] Adamsons, K., Blackman, G., Gregorovich, B., Lin, L., Matheson, R., Progress in organic coatings, 1998, 34, 64.

[4] Chiefari, J., Jeffery J., Mayadunne, R.T.A, Moad G., Rizzardo, E., Thang, S.G., Chain Transfer to Polymer: A convenient route to macromonomers, Macromolecules, 1999, 32, 7700.

[5] Callais, P.A., Moskal, M.G., Higher solids, higher quality acrylic coatings resins produced with organic peroxide initiators, Arkema Inc., King of Prussia, PA.

[6] Grady, M.C., Quan, C., Soroush, M. Thermally initiated polymerization process. US Patent Application Number 60/484,393, filed on July 2, 2003.

[7] Quan, C., Soroush, M., Grady, M.C., Hansen, J.E., Simonsick, W.J. Characterization of thermally polymerized n-butyl acrylate and ethyl acrylate, Macromolecules, 2005, 38(18), 7619.

[8] Rantow, F.S., Soroush, M., Grady, M.C., Kalfas, G.A., Spontaneous polymerization and chain microstructure evolution in high-temperature solution polymerization of n-butyl acrylate, Polymer, 2006, 47, 1423

[9] Pryor, W.A., Lasswell, L.D., Advances in free radical chemistry, Academic Press, New York, 1975, Vol. V.

[10] Mayo, F.R., Chain transfer in the polymerization of styrene. VIII. Chain transfer with bromobenzene and mechanism of thermal initiation, J. Am. Chem. Soc., 1953, 75, 6133.

[11] Khuong, K.S., H. Jones, W.H., Pryor, W.A., Houk, K.N. The Mechanism of self-initiated thermal polymerization of styrene. Theoretical solution of a classic problem, J. Am. Chem. Soc., 2005. 127, 1265-1277.

[12] Flory, P.J., The mechanism of vinyl polymerizations, J. Am. Chem. Soc., 1937, 59, 241.

[13] Hohenberg, P and Kohn, W, Inhomogeneous electron gas, Phys Rev., 1964, 136, 864.

[14] Salem, L., Rowland C., The electron properties of diradicals, Angew. Chem. Internat. Ed., 1972, 11, 92.

734g Computational Modeling of the Hydrolysis Reaction of Uranium Hexafluoride

Stephen L. Garrison and Jim M. Becnel. Savannah River National Laboratory, Savannah River Site, Bldg 773-42A, Rm 181, Aiken, SC 29803

Uranium hexafluoride (UF6) is synonymous with the nuclear industry as the basis for both the gaseous diffusion and gas centrifuge enrichment processes. The U.S. Department of Energy currently maintains approximately 57,000 cylinders containing a total of almost 700,000 metric tons of depleted UF6. Of special concern is the reaction of UF6 with water due to the corrosion caused by the hydrogen fluoride (HF) produced, which can significantly degrade the tanks commonly used for storage. Additionally, UF6 is highly toxic, and the thermochemistry of its hydrolysis is important to the environmental pathways of any accidentally released UF6.

This presentation will summarize results of a search for transitions states of the direct UF6-water gas-phase hydrolysis reaction using Density Functional Theory and small-core, relativistic pseudopotentials. No symmetric transition state leading to the direct formation of uranium oxide tetrafluoride, UOF4, was found, indicating that the gas-phase hydrolysis includes more steps than commonly mentioned mechanisms. One asymmetric transition state was found, and this appears to be the first published transition state for the important gas-phase reaction of UF6 with water.

734h First Principles-Based Atomistic Modeling of the Growth and Structure of Native Defects In Silicon

Sangheon Lee, Chemical Engineering, The University of Texas at Austin, 1 University Station Stop C0400, Austin, TX 78712 and Gyeong S. Hwang, Chemical Engineering, Dept. of Chem. Eng.,The University of Texas at Austin, 1 University Station Stop C0400, Austin, TX 78712.

Silicon self-interstitial defects have long been a subject of great interest because of their scientific and technological importance. Single Si interstitials are highly mobile even at room temperature. Hence, in bulk Si most Si interstitials are likely to remain in the form of self-interstitial defects or interstitial-impurity complexes. The formation and structure of rod-like {311} defects have been well characterized by high-resolution transmission electron microscopy. In addition, a series of recent spectroscopy measurements have evidenced existence of small self-interstitial clusters (below a few nanometers in equivalent diameter) before they evolve into larger extended defects. In ultrashallow junction formation with low-energy implanted dopants, such small interstitial clusters are thought to be a main source for free interstitials responsible for dopant transient enhanced diffusion and electrical deactivation during post-implantation thermal treatment. Hence, significant experimental and theoretical efforts have been made to determine the structure and stability of small self-interstitial clusters as well as their growth and structural transition to larger extended defects, yet still unclear. In this presentation, we will focus on addressing the evolution of Si self-interstitial clusters from compact to rod-like extended structures. Using a combination of continuous random network model based Metropolis Monte Carlo, tight-binding molecular dynamics and density functional theory calculations, we have determined the structure and formation energies of n-interstitial clusters ($n = 1 - 16$). The extensive combined calculations demonstrate a distinct structural transition from the compact to elongated chain-like shape at $n = 11 – 12$. In particular, the I12 structure elongated along the [110] direction turns out to be very stable with a formation energy of 1.65 eV per interstitial, substantially lower than 2.79 eV, 2.06 eV and 1.85 eV as predicted for I2, I3, and I4 clusters. Using tight binding molecular dynamics simulations we have also found that additional interstitials placed around the I12 cluster preferentially migrates to the (110) cluster edges to grow larger. In addition, the predicted formation energies in general decreases with cluster size, however the variation exhibits an oscillating trend, with strong minima when $n = 4$, 8 and 12 due to the high thermal stability of the I4, I8 and I12 clusters. Our results are overall in good agreement with earlier inverse model studies based on experimental observations. We will also discuss structural reconfiguration of small compact clusters ($n < 10$) and evolution of rod-like defects with a {311} habit plane from small clusters.

762a A Tailored Strategy for Pde-Based Design of Hierarchically Structured Porous Catalysts

Gang Wang, Department of Chemical and Biological Engineering, Rensselaer Polytechnic Institute, 110, 8th Street, Troy, NY 12180, Chris R. Kleijn, Department of Multi Scale Physics, Delft University of Technology, Prins Bernhardlaan 6, 2628 BW, Delft, Netherlands, and Marc-Olivier Coppens, Howard P. Isermann Department of Chemical Engineering, Rensselaer Polytechnic Institute, 110 8th Street, Troy, NY 12180.

Optimization problems involving the solution of partial differential equations (PDE) are often encountered in the context of optimal design, optimal control and parameter estimation. Based on the reduced gradient method, a general strategy was proposed to solve these problems by using mostly existing software. As an illustration, this strategy was employed to solve a PDE-based optimization problem that arises from the optimal design of the network of pore channels in hierarchically structured porous catalysts. A Fortran implementation was developed by combining a gradient-based optimization package, NLPQL, a multigrid solver, MGD9V, and a limited amount of in-house coding. The value and gradient of the objective function were computed by solving the discretized PDE and another system of linear equations using the MGD9V, and were fed into the NLPQL. The PDE was discretized in terms of a finite volume method on a matrix of computational cells. The number of cells ranged from 129x129 to 513x513, and the number of optimization variables from 41 to 201. Numerical tests were carried out on a Dell laptop with a 2.16 GHz Intel Core2 Duo processor. The results showed that the optimization typically converged within a limited number (i.e., 9-48) of iterations. The CPU time was from 2.52 to 211.52 seconds. The PDE was solved 36-201 times in each of the numerical tests. This study calls for the use of our strategy to solve PDE-based optimization problems.

762b A New Multiscale Coarse-Grain Methodology: The Self-Consistent Force-Matching Method

Jhih-Wei Chu, Chemical Engineering, University of California, Berkeley, 101A Gilman, UC Berkeley, Berkeley, CA 94720 and Hyung Min Cho, Chemical Engineering, UC Berkeley, 6 Gilman, UC Berkeley, Berkeley, CA 94720.

Force matching may be considered as a multiscale method that can be used to develop coarse-grained models directly from all-atom (or fine-grained) simulations. Recently, it has been shown that the force-matching equation is closely related to the Yvon-Born-Green (YBG) equation that describes the hierarchal connection between the structural correlation functions of simple liquids. In this work, we show that it is possible to extend the force-matching method to solve the YBG equation self-consistently. The resulting methodology may thus be employed to compute the underlying pair interactions that are required to give rise to a specified pair distribution. The implication of such capability in molecular design will be addressed. The SCFM is applied to develop the coarse-grained force fields of liquid water and liquid hexane from all-atom molecular dynamics simulations. By reproducing pair distribution functions, we found that the agreement of three-body correlation functions between all-atom and coarse-grained simulations is also improved. The SCFM method is contrasted with the commonly inverse Monte Carlo strategy in developing coarse-grain potential. The issues of transferability of coarse-grain force fields, the thermodynamic and dynamic consistency of coarse-grained simulations, and the integration of coarse-grained and all-atom models will also be discussed.

762c Multi-Scale Modeling of Electrodiffusion through Antimicrobial Peptide Pores

Dan S. Bolintineanu, Allison A. Langham, Abdallah Sayyed-Ahmad, H. Ted Davis, and Yiannis N. Kaznessis. Chemical Engineering and Materials Science, University of Minnesota, 421 Washington Ave SE, 151 Amundson Hall, Mailbox 16, Minneapolis, MN 55455

Protegrins are small, cationic antimicrobial peptides (AMPs) that are believed to act against bacterial infections by compromising the integrity of the bacterial cell membrane. Experimental evidence suggests that protegrins, as well as other AMPs, act cooperatively to form nanoscale pore structures that completely disrupt normal membrane function. We have employed molecular dynamics simulations to study the structural features of an octameric protegrin pore in an anionic lipid bilayer. Structures obtained from the simulations are used as input to a continuum electrodiffusion analysis of ion permeation (commonly referred to as Poisson-Nernst-Planck theory). The coupled steady-state Poisson-Nernst-Planck equations are solved numerically, using the geometry, charge density, and ion diffusivity profile obtained from the MD simulations. This multiscale approach allows for a quantitative analysis of ion transport through a protegrin pore in the biologically relevant non-equilibrium state (in particular, ion transport in the presence of an applied field, or in the presence of a concentration gradient). The relatively low computational cost of this approach allows us to explore numerous aspects of ion transport through a protegrin pore, including current-voltage relationships, the relative importance of geometry and charge distribution, ion types and diffusivities, and external conditions. We have additionally extended this model to approximately capture the time-dependent ion concentration profiles through an entire cell, as well as the decay of the cellular transmembrane potential. In this approach, the cell membrane is modeled as a homogeneous layer with properties derived from the small-scale, single pore structure. The permeation of ions through the cell membrane and the resulting transmembrane potential loss are believed to be significant events in the bactericidal action of all antimicrobial peptides.

762d Coarse-Grained Kinetic Monte Carlo Models: Applications to Membrane Receptor Dimerization

Stuart D. Collins, Chemical Engineering, University of Delaware, 150 Academy St, Newark, DE 19716, Abhijit Chatterjee, Theoretical Division, Los Alamos National Laboratory, T-12 MS B268, Los Alamos, NM 87545, and Dion Vlachos, Director of Center for Catalytic Science and Technology (CCST), University of Delaware, Newark, DE 19716.

The ErbB family of receptors is dysregulated in a number of cancers, and the signaling pathway of this receptor family is a critical target for several anti-cancer drugs. Therefore, a detailed understanding of the mechanisms of receptors activation is critical. However, despite a plethora of biochemical studies and single particle tracking experiments, the early molecular mechanisms involving epidermal growth factor (EGF) binding and EGF receptor (EGFR) dimerization are not as well understood. Due to the large disparity of time and length scales involved in receptor dimerization reactions, we adapt the coarse-grained Monte Carlo (CGMC) simulation framework to enable the simulation of in vivo receptor diffusion and dimerization.

On-lattice kinetic Monte Carlo (KMC) simulations have extensively been applied to numerous systems. However, their applicability is severely limited to relatively short time and length scales. Recently, the CGMC method was introduced to greatly expand the reach of the lattice KMC technique. Herein, we extend the previous spatial CGMC methods to multicomponent species and/or site types. Numerical examples are presented to demonstrate the method. Furthermore, we introduce the concept of homogenization at the stochastic level over all site types of a spatially coarse grained cell. Homogenization provides a novel coarsening of the number of processes, an important aspect for complex problems plagued by numerous microscopic processes (combinatorial complexity). As expected, the homogenized CGMC method outperforms the traditional KMC method on computational cost while retaining good accuracy.

Using the CGMC method, spatial modeling of ligand-mediated membrane receptor dimerization reaction dynamics was peformed. Furthermore, the simulations demonstrate the importance of spatial heterogeneity in membrane receptor localization. Mathematical models, especially one that takes into account spatial heterogeneity, show mechanistic understanding of receptor activation that may in turn enable improved future cancer treatments.

762e Equilibrium Path Sampling Study of Methane Diffusion In Natural Gas Hydrates

Baron Peters[1], Nils E. R. Zimmermann[2], Gregg T. Beckham[3], Jefferson W. Tester[4], and Bernhardt L. Trout[4]. (1) Chemical Engineering, UC Santa Barbara, Engineering II, Rm 3339, Santa Barbara, CA 93106, (2) Department of Chemical Engineering, Hamburg University of Technology, Eissendorfer Str. 38, Hamburg, 21073, Germany, (3) National Renewable Energy Lab, Golden, CO 80401, (4) Chemical Engineering Department, Massachusetts Institute of Technology, 77 Massachusetts Avenue, Room 66-454, Cambridge, MA 02139

Diffusivities in natural gas hydrates are important but elusive transport properties. We describe an efficient and versatile equilibrium path sampling (EPS) approach for computing free energies. We use EPS along with transmission coefficient calculations and kinetic monte carlo simulations, to estimate the methane diffusivity within a structure I hydrate crystal. The calculations support a water-vacancy assisted diffusion mechanism where the methane hops from an unoccupied "donor" cage to adjacent "acceptor" cage. Transmission coefficient calculations confirm features of the free energy surface and provide dynamically corrected rate constants. The rate constants from atomistic simulations are used to simulate self-diffusion with the Gillespie algorithm. Self-diffusion rates are lower than the Einstein estimate because of the lattice connectivity and methane's preference for large cages over small cages. From a computational perspective we demonstrate that EPS can compute free energies for a broader class of coordinates than umbrella sampling with molecular dynamics. From a technological perspective, we estimate an important diffusion constant that has been difficult to measure.

Web Page: www.engineering.ucsb.edu/~baronp/CODES/EQ-PATH-SAMPLING/manuscript.pdf

Modeling of Unit Operations by Molecular Simulations

D. Babic, A. Pereira Neto, A. Pfennig

AVT - Thermal Process Engineering,
RWTH Aachen University, Wüllnerstr. 5, D-52062 Aachen, Germany.
Corresponding author email address: andreas.pfennig@avt.rwth-aachen.de

Abstract

The behavior of separation columns can be simulated by molecular simulation, where the influence of all relevant elements of a separation stage is taken into account. In order to decrease the computational time the column is divided in separation stages, where the simulation of each stage is carried out on an individual computer and the number of stages can be set to any desired value. By performing a dimensional analysis the residence time of the molecules in the molecular simulation can be adjusted such that the simulated mass transfer corresponds to a macroscopic stage, i.e. the column simulation is also able to depict the influence of the residence time on the efficiency of separation stages.

Introduction

The properties and interactions of the molecules involved in chemical-engineering processes determine the behavior and thus the design of the used equipment. Nowadays it is still considered as impossible to bridge the length and time scales between molecules and technical equipment in only one or a few steps of model hierarchy. Nevertheless it is generally accepted that the major effects in chemical engineering facilities can be modeled just with equilibrium or rate-based approaches. These effects have their origin on molecular level and corresponding data can be obtained by only taking the vicinity of the interface into account, i.e. the simulation of a few hundred or thousand particles is sufficient.

In this work a simulation tool has been developed that is able to predict the behavior of entire separation columns based on molecular simulations. In principle any countercurrent separation equipment can be described, until today we focus on distillation and solvent extraction. The benefit of this approach is that the simulations are performed at the exact operating conditions, i.e. compositions, temperature, pressure and their gradients corresponding to the real apparatus. Thus it is not necessary to simulate data over the entire range of possible temperatures and compositions, where the data for the relevant conditions are then obtained afterwards by correlation or interpolation. Furthermore molecular simulations are able to predict multi-component effects of mass transfer as well as complex equilibrium data accurately and consistently.

Modeling

In the molecular approach the column is depicted by consecutive separation stages, i.e. the simulation of one separation stage is performed for two flowing phases in contact. Hereby the outflow of one stage is directly the inflow of the next stage with respect to material flow, composition, and temperature. In order to implement this concept in a first step the code of a molecular-simulation program has been modified as to allow different compartments and boundaries between them to be accounted for in a simulation, i.e. specific boundary properties can be applied to molecules that cross a border. By using different boundary properties all relevant elements of a separation process can be depicted on molecular level: partly penetrable walls simulate the behavior of sieve trays, heating and cooling regions are induced by influencing the molecular velocities or feed and product flows by inserting and removing particles from the simulation. In a first approach a counter flow had been induced by introducing gravity. It was already shown that with this setup the principle behavior of a distillation column can be simulated with less than a thousand particles [1]. The agreement with experimental data is excellent (Fig. 1). This approach has been refined and applied to solvent extraction.

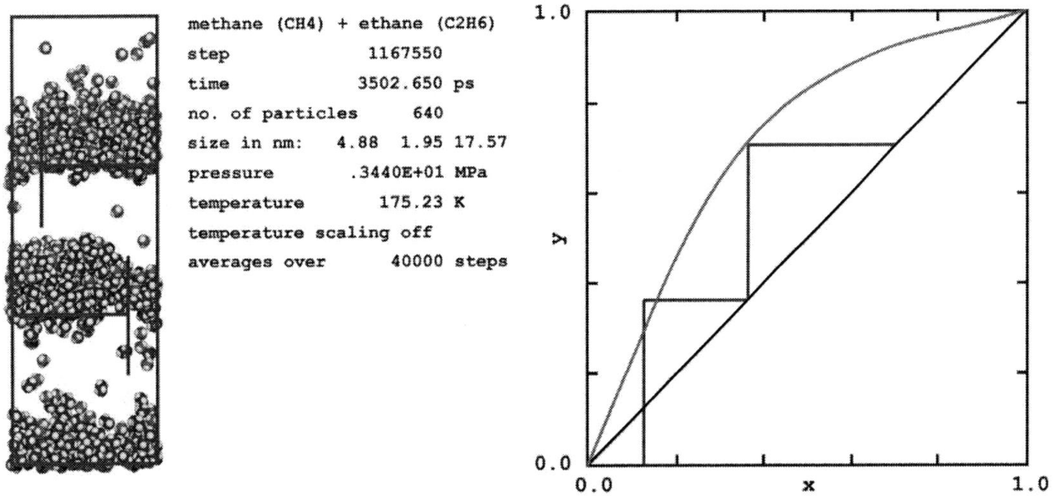

Figure 1. Snapshot of a distillation column simulated by molecular dynamics.

In order to make this method capable for liquid-liquid systems one has to be aware of the computational cost of the simulation presented in Fig. 1, which took about one week on a PC. Thus neither more particles, i.e. two liquid phases, nor more stages can be simulated with this method in reasonable simulation times. Therefore it is obvious that the method has to be parallelized. For that purpose the simulation cell has been cut into individual separation stages, where each stage can be simulated on an individual PC. By connecting the individual stages with their entering and leaving fluxes a separation column is depicted. The information that has to be exchanged between the stages, i.e. the PCs, doesn't have to include data regarding every individual molecule that left one stage but only overall data like the amount of substance, the composition or temperature of the flow rates are required. Additionally not after every simulation step the information has to be exchanged. It is sufficient to update the feed and product flows of every stage only after a few ten thousand steps. Thus the computational cost of the data exchange between the stages is negligible and the number of separation stages can be set to any desired value.

In the McCabe-Thiele plot on the right hand side of Fig. 1 one can see that the sieve trays of the column essentially behave as ideal separation stages, i.e. the products leaving one stage are in equilibrium. Due to the short length scales in the molecular simulation the residence time of the particles on one stage is too high, which means that equilibrium will be achieved. In order to simulate also the non-equilibrium character of separation stages on molecular level a dimensional analysis has been performed which provides the necessary residence time of the fluids in a molecular simulation that corresponds to a macroscopic separation stage.

In the triangular diagram in Fig. 2 the compositions of the feed (open circles), the products in the case of high (solid line) and low residence times (dashed line) and the equilibrium data of a separation stage simulated on molecular level are shown. The equilibrium data were obtained in simulations where the feed and product flow rates were set to zero, i.e. for infinite residence times.

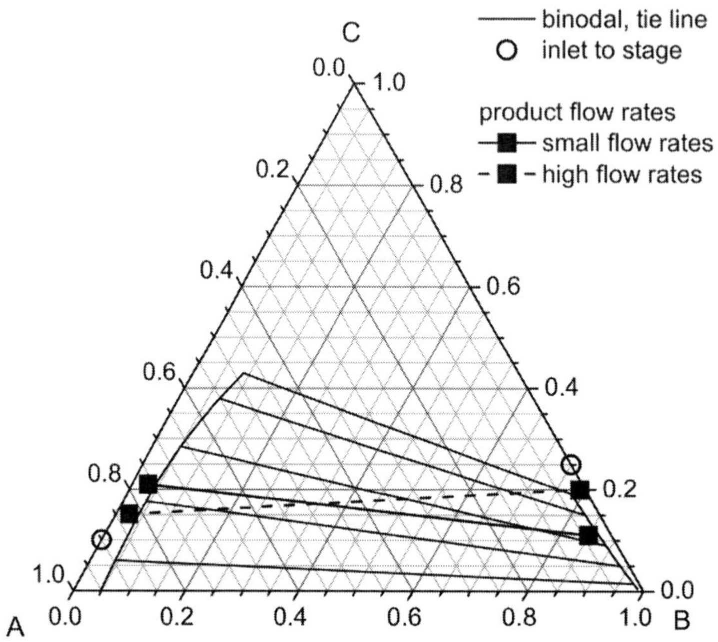

Figure 2. Compositions of streams leaving an extraction stage in the case of high (solid line) and low residence time (dashed line).

In both cases, short and long residence time, the A-rich product flow has a higher C concentration than the feed flow. Due to the finite residence time in both cases the products are not in equilibrium, i.e. their connecting line doesn't fall on a tie line. Fig. 2 also shows that increasing the flow rates (dashed line) leads to less C in the A-rich phase than in the case of low product flow rates (solid line), i.e. the product compositions are less equilibrated. Hence, with these molecular simulations it is possible to reproduce the influence of the residence time on the efficiency of a separation stage and the molecular simulation does not only depict adequately combined equilibrium stages like suggested by the McCabe-Thiele plot in Fig. 1.

In order to simulate an extraction column more than one separation stage has to be regarded. Therefore individual simulations as shown in Fig. 2 have to be connected, i.e. the outlet of one stage is the inlet of the consecutive stage and vice versa. In Fig. 3 a snapshot of such a simulation is presented. On the left hand side a schematic sketch of the column is shown, where the blue particles indicate component A, the red ones component B and green the transferred component C. The inlet and outlet streams are indicated by the arrows connecting the individual stages. On the right hand side the process is represented in the triangular diagram. The compositions of the feeds (F) are indicated by circles, the compositions of the products (P) by squares and the inlet and outlet fluxes connecting the individual stages by the dashed line.

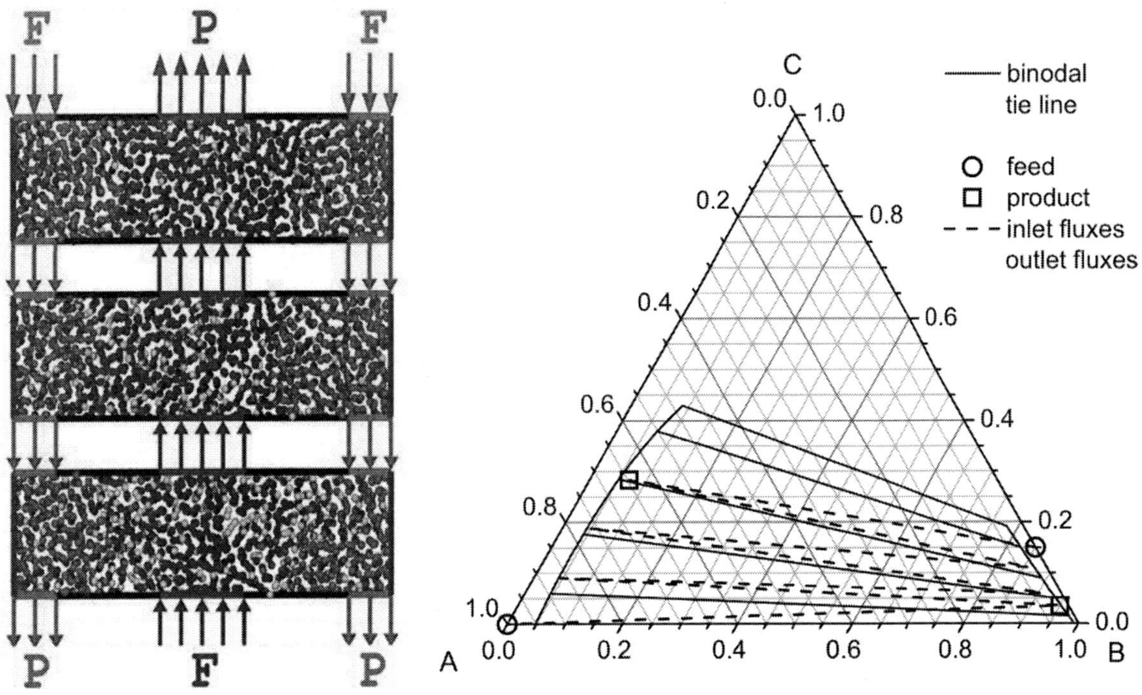

Figure 3. Simulation of an extraction column on molecular level.

Conclusion

The results in Figs. 1 and 3 demonstrate that a design of unit operations by molecular simulation is possible. To that end not all molecules in the column have to be regarded but only the relevant conditions have to be taken into account. This means that simulations that are performed in the vicinity of the interfacial region are sufficient to depict one separation stage. By connecting the inlet and outlet fluxes of the stages all relevant conditions are regarded that the particles pass in the column. But the simulations don't just behave like adequately connected equilibrium calculations. As can be seen in Fig. 3 the lines connecting the product compositions of one stage don't coincide with tie lines, i.e. the separation stages aren't ideal separation stages. They also include the non-equilibrium character of a stage, which can be either caused by multi-component effects of the mass transfer or by the finite residence time of the fluids on one stage. A dimensional analysis provides the necessary residence time of the fluids in the molecular simulation so that the mass transfer in the simulation corresponds to that of a macroscopic stage.

Currently we are comparing simulations with data from columns on pilot-plant scale. Also simulations with 20 and more stages are under way on a multi-CPU high-performance cluster.

References

1. A. Pfennig, "Distillation Simulated on Molecular Level," Mol. Sim. **30**(6), 361-366 (2004).

771a The Problem and the Benchmark Data

James Olson, Analytical Chemistry, The Diow Chemical Company, 770 Building, South Charleston, WV 25303

A description of the simulation challenge problem and the benchmark data used to judge it.

771b Place Holder for Talks by Entrants

Jonathan Moore, The Dow Chemical Company, 1702 Washington St., Midland, MI 48674

Presentations to be given by challenge participants

771c Announcement of the Champion and the Next Challenge

Jonathan Moore, The Dow Chemical Company, 1702 Washington St., Midland, MI 48674

The Champions of the 5th simulation challenge will be announced and the next simulation challenge discussed.

AUTHOR INDEX

AUTHOR INDEX

AUTHOR INDEX

AUTHOR INDEX